STORMS ABOVE THE DESERT

DATE			

STORMS ABOVE THE DESERT

Atmospheric Research in New Mexico
1935–1985

JOE CHEW
with the assistance of Jim Corey

Introduction by Bernard Vonnegut

Published in cooperation with the
Historical Society of New Mexico

UNIVERSITY OF NEW MEXICO PRESS
Albuquerque

Design: Milenda Nan Ok Lee

Library of Congress Cataloging-in-Publication Data

Chew, Joe.
  Storms above the desert.

  (Historical Society of New Mexico series)
  "Published in cooperation with the Historical
Society of New Mexico."
  Bibliography: p.
  Includes index.
  1. Thunderstorms—Research—New Mexico—South
Baldy Peak—History.   2. Atmospheric physics—
Research—New Mexico—South Baldy Peak—History.
3. Irving Langmuir Laboratory for Atmospheric
Research—History.   4. Meteorological stations—
New Mexico—South Baldy Peak—History.   I. Title.
II. Series.
QC968.C47   1987   551.5'0720789'91   87-6020
ISBN 0-8263-0983-6
ISBN 0-8263-0984-4 (pbk.)

To Ross Lomanitz,
and to college presidents
willing to take a chance.

Contents

Illustrations

Color Illustrations following page 94

Foreword

Storms Above the Desert is the fourteenth volume in the series copublished by the University of New Mexico Press and the Historical Society of New Mexico. This series of books brings to the interested public topics that might not otherwise be available, such as this present study. *Storms Above the Desert,* for example, delves into the realm of the history of science in New Mexico. Not only is this book the first in this series to deal with science, but *Storms Above the Desert* is also one of only a few book-length studies on scientific research in a state that for over forty years has been the site of investigations that are on the leading edge in their fields.

There are, of course, many individuals and organizations that devote themselves to the study of the history of science. The focus of this book is the Langmuir Atmospheric Laboratory on the crest of the Magdalena Mountains some twenty miles west of Socorro, New Mexico. At first brush this would seem a very limited subject indeed; however, as so often happens in science, the history of an institution and the people associated with it reflect major currents in scientific debate as well as concerns of society in general. In *Storms Above the Desert,* a study of the Langmuir Atmospheric Laboratory is inseparable from the broader areas of thunderstorm research and weather modification. As this book makes clear, weather modification has general application to economic, social, and even political issues. This book develops these larger themes as a background to set the stage for the particulars of Langmuir and the scientists who continue to work there.

The authors did not content themselves with the narrative history of the founding of the Langmuir Laboratory. They delve into the technical matters of scientific research, but they do so in a most engaging and easily understood account. While the scientific debates associated with weather modification and atmospheric research are beyond the ken of many historians and nonspecialists, the authors and researchers of this volume are trained to write and also have the technical education to address the subject. This book is a fine example of the contribution practitioners of the Technical Communications degree offered at New Mexico Tech can make in presenting scientific work that is accessible to all readers. Professor James Corey and his students, and in particular Joe Chew, present in *Storms Above the Desert* an authoritative and captivating history of the atmospheric research conducted in New Mexico since 1935. The members of the Publication Committee of the Historical Society Board are delighted to sponsor this book.

The Board of Directors of the Historical Society of New Mexico is made up of interested citizens and representatives from the academic community. The current officers and members of the board are: Spencer Wilson, president; Charles Bennett, 1st vice president; Michael L. Olsen, 2nd vice president; John W. Grassham, secretary; M. M. Bloom, Jr., treasurer; and Carol Cellucci, executive director. The members of the board are: John P. Conron, Thomas E. Chavez, Richard N. Ellis, Austin Hoover, John P. Wilson, Albert H. Schroeder, Loraine Lavender, William J. Lock, Octavia Fellin, Myra Ellen Jenkins, Susan Berry, Darlis Miller, Morgan Nelson, Robert R. White, Robert J. Torrez, and Elvis E. Fleming.

Introduction

Most people I talk to are surprised to learn how little scientists know about thunderstorms. Experts don't really understand the complicated motions of the updrafts and downdrafts, how rain and hail form, or the various processes that are responsible for generating the electrical energy that causes lightning.

It is easy to understand why people assume everything is known about thunderstorms. They have captured man's attention from earliest times, and were among the first natural phenomena that early scientific pioneers, such as Lucretius, Franklin, and Lomonosov, attempted to explain. One might suppose that by now, after several hundred years, we would have answered most of the fascinating questions thunderstorms pose, but we have not.

How little we know about thunderstorms is even more surprising in view of the important role they play in our lives. The winds, hail, lightning, floods, and tornadoes that thunderclouds produce can threaten us, take many lives, and cause great damage. But even more importantly, thunderclouds are beneficial. They provide most of the water that is required to grow the world's food. It is, therefore, not only to satisfy our curiosity, but very much in our own interest that we strive to learn how they work.

Thunderstorms have proven remarkably resistant to scientific investigation. In contrast to many phenomena that are continuously or predictably available for study, the occurrence of thunderstorms is highly variable in space and time. A great deal of skill and luck is required just to be in the right place at the right time to make observations. But being there is only the beginning; it is extraordinarily difficult and often dangerous to make measurements within or near an active storm with its violent turbulence, lightning, and hail.

Now, with modern technology, tools are becoming available that will enable scientists to secure the information necessary to understand the various processes taking place in the clouds. Among the leaders in thunderstorm research, nationally and internationally, are the scientists at the New Mexico Institute of Mining and Technology. It is fortunate that the twenty-fifth anniversary of Langmuir Laboratory has provided the occasion to recognize the achievements of this group.

Joe Chew, his collaborators, and their mentor Jim Corey are to be congratulated for providing an accurate and detailed record of the establishment of this facility and the steady stream of important contributions that have come from the scientists working there. Storms above the Desert will be an important chapter in the scientific history of thunderstorm investigation. At the same time it will provide nonspecialized readers with new insights into how the fascinating human endeavor we call scientific research comes about.

This account is particularly interesting to us because it tells much about the origins of Langmuir Laboratory that I had not known before and concerns close personal friends responsible for organizing and carrying out the unique and unusually productive research conducted there. I count myself as exceedingly fortunate that my interests in cloud seeding and atmospheric electricity have brought about my long and close association with the scientists at New Mexico Tech and Langmuir Laboratory.

Bernard Vonnegut
Atmospheric Sciences Research Center
State University of New York at Albany

Preface

Storms Above the Desert came into existence in an unusual way.

It was planned and researched by a college writing class at the New Mexico Institute of Mining and Technology. That class had twelve juniors and seniors enrolled for the fall 1984 term. I assigned them to writing the history of Langmuir Laboratory and thunderstorm research in New Mexico as an exercise in communication processes: organizing, researching, interviewing, writing, and editing. In other words, the collaboration and teamwork, as well as the writing skills, they would use later in industry. I was most gratified as a teacher to find that from the class project came a publishable manuscript.

Storms above the Desert was written by one of those twelve students: Joe Chew. Taking the raw material provided by his classmates, he transformed it into a coherent, interesting historical narrative. That Joe would be the author of record was no accident. Early in the project, it became clear that he possessed both the knowledge of the subject and the maturity of style to write the final draft of the book. He took control of the manuscript at an early stage and stayed with it far beyond the end of the fall term. It truly is his creation.

The publication of the book was made possible through a special program of the Historical Society of New Mexico. In association with the University of New Mexico Press, the Society periodically elects to underwrite the publication costs of manuscripts that, it feels, preserve significant parts of New Mexico's past. We are most grateful to the Society for selecting Storms Above the Desert.

Then there is the subject itself.

The year 1985 was the silver anniversary of the decision to fund and build what is now the Irving Langmuir Laboratory for Atmospheric Research, a thunderstorm laboratory high in the Magdalena Mountains of west-central New Mexico. Run by New Mexico Tech (formerly the New Mexico School of Mines), it is the only laboratory of its kind in the United States and one of the very few in the world. The work done there each summer during the all-too-brief thunderstorm season has resulted in some major contributions to the understanding of thunderstorms and lightning.

Coincidentally, 1985 was also the fiftieth anniversary of atmospheric research in New Mexico. The science of atmospheric physics was in its infancy when Dr. E. J. Workman began studying thunderstorms at the University of New Mexico in 1935. From those depression-era beginnings rose not only pure science, but also the art of rainmaking and other practical things. Although weather modification is out of fashion these days, atmospheric research is doing better than ever. Some of the top experts in the field believe that they are on the verge of truly understanding just what goes on inside a thunderstorm. But, then again, people have said that before.

Those who imagine science to be an abstracted, impersonal pursuit will be surprised to learn how atmospheric physics was torn and reshaped by academic politics, personality conflicts, and even a lawsuit or two. The fifty-year saga of thunderstorm research has been a human drama as much as a scientific one; the science itself, though, is unusually interesting. We hope that we have accurately portrayed the people who played roles in that half-century, and, even more, that we have also done justice to the science.

A great many people have assisted directly and indirectly in this project. Noteworthy contributions were made by the students of TC 301, in particular Kim Eiland, Toni Ball

Johnson, and Kathy Smith, and also including Jill Bartel, Chris Benedict, Rick Clyne, Diane Hattler, Terry Jackson, Mary McClure, Heidi Miller, and Dave Pellatz. Special thanks are due Dr. John DeWitt McKee, Professor Emeritus of English at New Mexico Tech, whose suggestions on style greatly improved the quality of the manuscript; to Dr. Paige Christiansen, Professor of History, who provided valuable leads for fruitful research, and to Dr. Spencer Wilson, Professor of History, who helped secure a publisher for the manuscript. The support of Dr. Marx Brook, who gave us access to his photo archives, and of Floyd Willard, who runs the New Mexico Tech Research and Development Division photo lab, was invaluable in helping us illustrate the book. Beverly Ohline, former director of Tech's Information Services Office, was very helpful in providing news releases, photographs, and facilities, and the Tech librarians were similarly helpful when it came to providing research materials from the school's archives.

But we would especially like to thank the scientists we worked with: Bernard Vonnegut of the State University of New York; Robert Holzer of the University of California at Los Angeles; New Mexico State Engineer Steve Reynolds; and New Mexico Tech's Marx Brook, Charles Holmes, William Hume, Dan Jones, Paul Krehbiel, Charles Moore, Marvin Wilkening, and William Winn. By cooperating with our historical research and by figuring importantly in the story itself, all of them have made this book possible.

Jim Corey

Langmuir Laboratory on remote South Baldy. New Mexico Tech Archives.

1

The View From A Height

The ridge extending from South Baldy, a 10,700-foot peak in New Mexico's Cibola National Forest, is about the last place anyone would look for a five-story building. But there it is: green, rectangular, and bristling with antennas, overlooking the seven miles of primitive road leading up to it. Next door is another building, a round one, topped by what looks like an airport control tower.

The rectangular building, the Annex of the Langmuir Laboratory for Atmospheric Research, also has an observation tower on top. In the early morning sun, both towers offer a fine view of the white antennas of the Very Large Array radiotelescope spread out in a giant Y, thirteen miles on a side, across the Plains of San Agustin to the west. By night, Albuquerque is a lake of golden light in the middle distance, Santa Fe a glittering jewel on the northern horizon. The light spills down the Rio Grande Valley to Socorro and trickles away into the desert to the south.

But on this August afternoon, no one at Langmuir is looking at the scenery. Scientists and their assistants are finishing their lunches and looking for raincoats and walkie-talkies. The cumulonimbus cloud overhead is getting bigger and darker, signaling that is it time to go to work.

On a chart recorder in the observation tower atop the Annex, the fair-weather meanderings of the electric-field readings cross the center lines of their tracks, smoothing out and veering off into the foul-weather zone. In a blacked-out room one floor below, a student watches a pair of green radarscopes, his face bathed in their eerie glow as he monitors the growth of the storm. At the Socorro airport, seventeen miles away, a powered sailplane takes off on a mission to penetrate the cloud.

The stately tick of the chart recorders becomes a rattle as the tower operator puts them on high speed. Along the ridge, activity grows like the cloud itself. More radars are warmed up and put in operation. A tethered balloon carries instruments into the cloudbase. As the first raindrops streak the windows, the plane buzzes down Sawmill Canyon, climbing past the lab into the cloud.

Two-way radios crackle with reports on the storm's growth as the rain falls harder and the electric fields become stronger. At the very top of South Baldy, the metal doors of two underground steel bunkers—Kivas I and II—are closed. Inside the kivas, scientists check their instruments and wait for the electric fields to build even higher. An array of small rockets stands ready atop one of the kivas.

The electric fields grow: six kilovolts per meter (kV/M, pronounced kay-vee-em) then seven, then eight. A man in his sixties with crewcut and green jumpsuit climbs into the tower. Energized by the excitement, he picks up the radio mike and reports personally on what the radarscopes revealed.

Eight point five kVm. The plane is safely away from the lab. The ground personnel are at their posts. Radars and cameras are trained on the peak as the fields pass nine kV/m and a woman's voice comes over the radio: "Langmuir Lab, this is the Kiva. Request a five-minute window for firing."

Crewcut nods his assent. "Five-minute window open 1335," replies the tower operator, who gives a warning blast on a siren. A lightning strike on Timber Ridge, on the other side of the canyon, discharges the fields, which quickly build again: nine kV/m, ten, ten five. A potential difference of ten thousand five hundred volts between the floor and a person's belt buckle.

"Langmuir Lab, we will fire in five seconds.

Closeup of laboratory. Jack B. Pearce photograph.

"Four.
"Three.
"Two.
"One.
"Fire!"

A streak of black smoke arcs into the murk from the top of the Kiva. Just as the hollow whoosh of the rocket reaches the lab, a column of light streaks from the cloud to the Kiva. The crack and roll of thunder mingles with the whoops and yells of onlookers as the operator confirms the triggered stroke.

To be sure, thunderstorm research means solving equations, interpreting data, and looking up references in the library. But during the few weeks of storm season at Langmuir, it means other things. It means standing on the roof while corona current sings from the antennas, then hustling back into the tower seconds before lightning strikes so close there is no thunder. It means firing a rocket from the Kiva and waiting underground for tens of thousands of amperes—and only God knows how many millions of volts—to go to ground through the steel roof. And it also means nights of quiet solitude for those who live at the lab, with the rest of the world a glittering insignificance far below.

Some people don't last a single summer at the lab; others build their careers there.

In many ways, the science of atmospheric physics, which gave birth to this lab on top of a mountain, can be traced back to Benjamin Franklin. Some of its more fanciful practitioners look back even further, to the Roman Lucretius, or even to the Book of Exodus. But many atmospheric physicists—certainly the ones in New Mexico—like to place the beginnings of their discipline in 1935, when a physicist named E. J. Workman began his studies of clouds and lightning. The lab itself got its start in 1961, largely because of the efforts of Dr. Workman.

2

The Early Days

On March 23, 1961, not long after the unheralded silver anniversary of atmospheric physics, a group of men sat in a meeting room at New Mexico Tech in Socorro. Six of them were from the Advisory Panel on Atmospheric Sciences of the National Science Foundation, or NSF; seven more were drawn from the school's atmospheric physics researchers. Tech needed a lot of money to build a mountain lab, and the NSF had it.

This undercurrent of money ran throughout a discussion of atmospheric research. Tech was one of the pioneering institutions in the field. The school's president, E. J. Workman, had come there just after World War II, back when it was still the New Mexico School of Mines, bringing the beginnings of atmospheric physics with him. The study of thunderstorms, born in the Depression and reborn after the war, had gone through some lean years, but now the field was prospering—thanks in part to money from the NSF.

Although Tech was acquiring a name in atmospheric research, the work had never had a single, focal research facility. Trailers scattered like buckshot north, south, and west of Socorro served as scientific worksites. In results the work was up-to-date, but in spirit it still recalled the days when Workman had chased thunderstorms with an instrument-carrying Packard roadster. For several years, he and his colleagues had wanted to build a mountaintop lab: there would gather all the researchers, and it was to be located where storms would come often.

Now there was a chance to get the money—and never in his career had Workman been accused of passing opportunity's door without knocking. Out of a brief discussion in a spartan meeting room came a $200,000 grant, followed by $300,000 the next year, to build the Irving Langmuir Laboratory for Atmospheric Research.

On the day they were talked into honoring the school's standing request for a building on top of a mountain, the NSF representatives were negotiating in Workman's territory. In fact, they were practically negotiating in his living room. The elderly physicist lived in the research building where they were gathered. His living quarters were upstairs, next to his research tower. Downstairs was his office as president of the tiny college. By day he ran the school and worked in his lab. By night he worked in the lab or wandered the halls, thinking his thoughts and turning off the lights in empty rooms. Perhaps he was lonely; perhaps he was merely busy. But he was undoubtedly at home. For a quarter of a century his name had been synonymous with lightning research in New Mexico, and he had done much of that work while president of the school in Socorro.

In 1961, Everly John Workman was sixty-two years old and had been president of New Mexico Tech for fifteen years. The school had virtually disappeared during World War II; for a long time thereafter it was what Workman made it. Nothing was too big for him to tackle, and nothing was too small to attract his attention. He had personally designed both the school's basic curriculum and the first nine holes of its golf course. He had taken the Atomic Energy Commission to court and won. And if a construction worker laid a crooked tile or hung a sticking door, he, President Workman, personally wrote it up and demanded that it be set right.

Colleagues, employees, even innocent bystanders were swept up by this one-man whirlwind. Workman was disliked by many people, but he was scorned by no one. That much energy inspires a certain hushed respect, even in one's enemies.

E. J. Workman, on a visit to the lab in 1979, with student technician Anne Myers. New Mexico Tech Information Services.

8

Workman first came to New Mexico in 1933. He was then at the University of New Mexico in Albuquerque, an institution with departments the size of the entire School of Mines. "Jack" Workman, a thirty-six-year-old journeyman researcher with a growing reputation in his field, became the head of the physics department during his third year at UNM. Then, teaming up with Robert Holzer, a member of his faculty, Workman organized a whole new arm of the university: the Research and Development Division.

That move sounds more dramatic than it actually was. The tiny new division had a dynamic leader, a sharp junior partner, and a thoroughly grandiose name. The only thing it did not have was money. Workman's budget for equipment and supplies one year was $300. The Virginia Academy of Science took pity on him and chipped in another $50.

The brand-new Research and Development Division began in poverty and was obviously going to stay there for some time. Workman pragmatically looked for something that could be researched and developed for $350. Looking up to the skies in desperation, he found the thunderstorm.

Albuquerque just happened to be a great place to study thunderstorms. The clouds that dot the skies of New Mexico on hot summer days tend to be small, isolated, and relatively stationary. They are born over solar-heated local terrain and move only a few miles before dying. These characteristics make them easier to study than the big frontal storms in the congested skies of the East.

More pertinent to Workman was the fact that storms were an ideal subject for the R&DD to study during the Depression. Storm behavior was a young, wide-open field of inquiry, and the pioneering exploration of it could be done for practically nothing. For all intents and purposes, Workman was founding his own science. That meant building a foundation for later work by observing in detail exactly what happens during the life cycle of a lightning stroke or thunderstorm.

Fortunately, this work did not require expensive technology. Want to know the strength of the updrafts in a convective cloud turret? Get a stopwatch and a theodolite, sit out in the desert, and measure the rise and fall of the cloudtops. Need to find out what steps are involved in a lightning stroke? Build a camera that has its film mounted on a rotating disk so you can get side-by-side images of the successive stages. Want a real bargain in scientific knowledge? Sit down with pencil and paper, and try to use these data to figure out how a cloud might work.

Some researchers have pointed out that atmospheric research was evolving through the same stages that medicine had passed through several centuries earlier. Workman and Holzer were studying the anatomy of the thunderstorm, identifying the phases of its life cycle, noting the roles of different parts. The physiology of the storm could be studied only after the anatomy was known. While learning the anatomy, they found, as is usually the case with basic research, that each answer they found gave birth to two or three more questions.

Workman and Holzer continued in this way for six years. They watched thunderstorms and swapped photos of unusual lightning strokes with other researchers. They wondered how long it would be until they understood exactly what was going on above their heads. After all, how complicated could a thunderstorm be? In those early years, Workman and his colleagues amassed a body of basic knowledge that would stand them in good stead later on.

As their knowledge and budgets grew, they turned to more-advanced problems. Then as now, one of the most tantalizing puzzles in atmospheric physics was the question of how clouds become electrified. Workman and his colleagues set up electric-field meters in the desert. When thunder-storms passed overhead, the readings told them, more or less, where the charging mechanism was located in the cloud. They figured out that the centers of charge were high enough to involve ice crystals and supercooled water droplets, a finding that would become very important.

That discovery led to more questions. How do the ice crystals become electrified? What is the nature of the particles that make up a cloud? Could such knowledge be put to a practical use? The scientists paid a local garage to weld a steel roof onto an open Packard and outfitted the car with instruments. In this lightning-proof rolling laboratory, they pursued thunderstorms along the primitive roads of New Mexico. (Workman, an atrocious driver, later outfitted his own car with various sensors. With the accelerator firmly planted to the floor, he would duck under the dashboard to check the readings on a chart recorder. His passengers just gaped, awaiting their doom, as the blurred scenery sped by.)

But a few months after the thunderstorm season of 1941, the scientists' days of cloud-watching in the desert came to an abrupt halt. The Research and Development Division found itself in military service. It was a rude interruption to the project, but both the R&DD and atmospheric physics in general would benefit from it.

In World War II, the military looked upon the wide-open wastelands of New Mexico and saw paradise. In the mountains north of Santa Fe a private boys' school, named Los Alamos, was selected as the intellectual center of the Manhattan Project. Huge volumes of mail suddenly pouring into a place as small as Los Alamos would have given away the secret location, so Workman's Physics Department address was used as a mail drop for the atomic bomb project. But the R&DD itself, eighty miles from the excitement of the Manhattan Engineering District, was put to work on a less-exotic device for which there was a pressing need: the proximity fuse.

Ordnance designers had known for a long time how to make bombs explode on impact. They also knew how to make time-delay fuses that would detonate artillery shells after a preset time of flight. Until World War II, impact detonation served well enough for attacking ground targets, and time-delay shells were thought to be good enough for anti-aircraft use.

World War II brought the art of using explosives to a new level of sophistication. Shrapnel rounds are most effective against personnel if detonated a few feet above the ground. Anti-aircraft shells don't kill reliably unless they are set off close to the airplane. Bombs destroy certain types of structures more easily if exploded nearby, thus setting up a shock wave. Time-delay detonation was no longer accurate enough. Instead, the military needed a fuse that could actually sense the distance to the target. The Federal Office of Scientific Research and Development called on Workman to help invent one.

The proximity fuse, a great advance in weapons technology, was basically a primitive pulse radar. Implanted in the nose of a shell or bomb, it shot radio waves at the target and measured the time lapse until the waves bounced back. When that time was short enough, it detonated.

The theory was simple enough. The problems were all practical: how to make a rugged, reliable, easy-to-manufacture proximity fuse that would fit into the fuse-hole in a shell. Workman spent most of the war on the project, shuttling back and forth between Albuquerque and Washington, D.C. As improved designs were introduced during the war, Workman specialized in testing them. For a time, the tallest wooden towers in the world stood on the desert between Albuquerque and the Sandia Mountains. Workman and his crew suspended airplanes between the two towers to serve as targets for proximity-fused shells.

They proved that the fuses worked as claimed, firing round after round to alleviate the peculiar angst of artillerymen, who fear that a new, untried device may go off inside the gun. They were less successful in dealing with a fear that haunted Army security men—the specter of an unexploded shell landing behind enemy lines with the precious secret still humming away in its nose. The Navy, untroubled by that worry, began using proximity fuses against Japanese aircraft in 1943. The Army finally had to break out the fuses in 1944 to defend against V-1 buzz bombs. Later that year, proximity-fused shells were used with great success by the Navy against Japanese kamikaze planes and by the Army against German troops in the Battle of the Bulge.

Most of the proximity-fuse work was done clandestinely, but the secret was out by the end of the war. In a June 23, 1947, ceremony at Los Alamos, General Omar Bradley personally commended Workman and R&DD researcher John S. Reinhart for their efforts. "We can go a long way," said Bradley, "if we develop weapons so terrible that we won't have to use them, and thereby make war farther away." (Even as Bradley let slip his unintentional double entendre, the nearby White Sands Proving Range was working with captured German V-2 rockets and American copies.)

Thanks to military research, the R&DD was off and running by 1946. Workman's shoestring operation had grown into a 200-member department. In addition to testing proximity fuses, the division had studied such things as high-velocity projectiles for penetrating tank armor and had tested the combat survivability of the new B-29 bomber. Meanwhile, thunderstorm research had not been entirely forgotten; since lightning can interfere with radio communication, the War Department had funded some investigations. But that work had been sporadic and limited in scope. When his war duties were over, Workman was eager to return to his primary interest, the thunderstorm.

But by 1946 the UNM president under whom the R&DD had grown and prospered was dead, and his replacement wanted a piece of Workman's action. Lucrative Navy and Signal Corps contracts were being administered personally by Workman. The new president wanted the money to go through the university, which would then take some of it. Today it is standard procedure for universities to take as much as 50 percent off the top of every research contract. Upwardly mobile professors try to build reputations for securing outside funding, and administrators look with favor on the ones who bring in the biggest grants and contracts. But in 1946, Workman would have none of that.

The UNM president laid down the law: Either the contracts go through the school, Jack, or there will be no contracts. Fine, replied Workman. I'll move to the School of Mines and take my contracts and my people with me! The president called his bluff, only to find that Workman wasn't bluffing. Governor John Dempsey called a special meeting of the regents of both schools to discuss the matter. In a single afternoon, Workman and most of his key people resigned from UNM and were hired by the School of Mines.

When Workman went south to Socorro, he took all of the R&DD research contracts with him, an action that slashed the financial resources of UNM. Along with him came about half the people on the R&DD staff and in the UNM Physics Department. The only catch was that there was no place for them to go. The School of Mines had only 111 students in 1946, and the small town of Socorro had no place for the huge influx of scientists to live. So Workman, adding insult to injury, leased the buildings of the defunct Sandia Girls' School in Albuquerque and kept the R&DD right in UNM's back yard.

Soon, more presidential trouble had Workman on the move again. The president of the School of Mines resigned for personal reasons, and the school's regents asked Workman to assume the interim presidency. He grudgingly agreed to take the job temporarily, while searching for his own replacement. Meanwhile, he lived in Albuquerque and commuted seventy miles to Socorro when necessary.

At the time, Socorro already had an absentee mayor; the prospect of also having an absentee president at the School of Mines was not well received by the townspeople. Soon, though, Workman's deeds made everyone angry enough to forget such trifles.

One of Workman's trusted administrative troubleshooters in the R&DD was Bill Glance, a retired Marine colonel. The newly chosen temporary absentee president sent Glance down to Socorro with orders to slim down the school's grossly oversized Physical Plant Department. Eleven employees (one for every ten students, and about one for every tree on the campus) found themselves out on the street. This provoked special dismay because they had gotten their jobs through local political patronage and had never dreamed that a grim-faced gringo from Albuquerque would send them packing.

Workman's next step was to disband the School of Mines basketball team. The team was a thoroughly mercenary outfit with a seven-foot player and a good record of tournament wins. A few of the players stayed on after the team was broken up, but three of them left the following year in a cheating scandal. The school never fielded another varsity team in any sport.

The radar convoy leaves the Sandia Girls' School for the move to Socorro, 1949. New Mexico Tech Archives.

Having outraged the community, Workman tilted his lance at the faculty. The School of Mines had been, for all practical purposes, a trade school for geologists and mining engineers. Workman, who knew quite well how fast science and technology were progressing, thought the curriculum would have to be broader, deeper, and harder to prepare students adequately for the future. He decided, for example, that calculus should be a requirement for graduation in every field. That was such a heresy in 1946 that three different state agencies investigated his professional competence, but, as usual, he got his way in the end.

The calculus requirement was just the first step in a stem-to-stern redesign of the curriculum (to Workman's specifications, of course.) Since not all of the old faculty members agreed with his plans, he had to throw some of them out— in one case, literally.

After three years, Workman's interim presidency was beginning to look permanent, but he was still commuting between Albuquerque and Socorro. The expiration in 1949 of the lease on the former Girls' School forced his hand. The Atomic Energy Commission wanted to condemn the buildings for what would become Sandia National Laboratories, so it pulled rank on Workman, not realizing that he would fight back in court. Although Workman lost the suit to keep the buildings, he forced the AEC to pay the R&DD more than $600,000 for the privilege. Since the R&DD owed only $300,000 or so for the facilities, the division had a substantial bank account when it loaded its instruments on trucks and headed for Socorro.

3

Tampering With The Weather

While Workman was reluctantly becoming a college president, scientists in a corporate laboratory half a continent away were preparing to modify the weather.

Before July 1946, the famous observation that "everybody talks about the weather, but nobody does anything about it" might as well have been a law of physics. Weather modification belonged in the realm of hokum until that July. Then, without warning or expectation, chance favored the prepared mind of a General Electric Laboratories scientist named Vincent Schaefer, and cloud seeding was born.

Although Schaefer's discovery was sudden and serendipitous, much of the research behind it had been going on at GE since 1940. In that year, Schaefer's boss, Dr. Irving Langmuir, was approached by the U. S. Chemical Warfare Service to study the filtration processes used in gas masks. That entailed generating various kinds of smoke and studying their properties on the particle level. Langmuir, a Nobel laureate and, by all accounts, one of GE's resident geniuses, had in Schaefer a hard-working associate who contributed greatly to the research. (Schaefer started his GE career as a machinist; after working with Dr. Langmuir, he came to be a respected scientist in his own right.)

Langmuir and Schaefer continued with this and other aerosol-related war research until 1943, when they began investigating the properties of supercooled clouds as part of a study of precipitation static. This led them into the problem of aircraft icing, which was especially timely; large numbers of aircraft were flying in icing conditions on missions that could not be delayed or aborted because of bad weather. They studied the problem on Mount Washington, New Hampshire, which experiences some of the worst winter weather in the world.

When he was not shivering atop Mount Washington, Langmuir made extensive calculations describing ice accumulation on various geometric shapes and materials. Combining these theoretical results with field observations, the researchers incidentally obtained a good working knowledge of cloud structures and the growth of cloud particles. Their work led them to study exactly what happens when water freezes. Schaefer needed to produce cloud-like conditions in the lab instead of seeking them in winter on a frozen, windswept mountaintop, so in 1946 he invented the "cold box," a home freezer—GE, of course—with a black velvet lining and a viewing light.

By breathing into the cold box, he could produce a little cloud of water droplets that condensed from his breath. Like those in the higher reaches of a cloud, these droplets were supercooled—below the freezing point, but still liquid. Being able to recreate these cloudlike conditions in the lab was a major practical step in cloud investigations. GE's public-relations people were almost as proud of it as he was (although they were exasperated by reporters who insisted on calling his freezer a Frigidaire.) In spite of the nomenclature dispute, this elegantly simple piece of equipment became a cornerstone of cloud research.

On July 13, 1946, Schaefer came to work and found that his cold box, usually kept at minus twenty-three degrees Celsius, had been turned off. Eager to recool it in a hurry, he went down the hall and got some dry ice. When he put the dry ice into the cold box, an abundance of ice crystals suddenly appeared in the fog. He experimented further and found that even a tiny grain of dry ice would produce this effect. (Later research showed that any substance with a temperature of minus forty degrees Celsius or below would do the job.)

Langmuir, who had been in California when Schaefer made the discovery, followed up with a theoretical study of the growth rate of ice nuclei produced by dry ice, calculating the velocity, fall time, and dissipation rate of the ice particles. When he saw how difficult and time-consuming this would be, he sought and found a junior member for the research team.

The man he found was Dr. Bernard Vonnegut, then in his first year with GE. He, too, had spent some of the war years studying airplane icing and had also worked on smoke filtration. His graduate work in physical chemistry had concerned the freezing points of very dilute aqueous solutions of various chemicals. As Vonnegut worked on the calculations with Langmuir, he realized that it might be possible to cause ice crystals to form in the cold box around particles whose crystal structure was similar to that of ice. These "ice nuclei," he hypothesized, might provide a kind of pattern or template on which water molecules would deposit themselves in the ice crystal arrangement.

The obvious next step was a systematic search for such nuclei. It was, at that time, a purely scientific investigation rather than a deliberate attempt to develop a new technique. After contemplating the crystal structures of more than a thousand substances, Vonnegut concluded that antimony metal and the iodides of lead and silver were the best candidates. At first, experimenting with various forms of these substances, he achieved little success—probably because (as he found out later) his bottle of silver iodide was heavily contaminated with sodium nitrate, a fairly good antifreeze. But on November 14, 1946, he introduced a smoke made of silver iodide particles into Schaefer's cold box and observed a spectacular fallout of ice crystals.

Irving Langmuir with camera. New Mexico Tech Archives.

In the meantime, Schaefer had become the first person to seed a cloud. On November 13, Schaefer and pilot Curtis Talbot successfully used dry ice to induce precipitation in a cloud—"an unsuspecting cloud over the Adirondacks," as Schaefer put it in a technical report. Having caught the four-mile-long cloud unawares, he and Talbot proceeded to plow a trough along its top with particles of dry ice. Snow began to fall from the cloudbase. Although the snow melted and evaporated before it hit the ground, the results were dramatic enough to change cloud seeding from a laboratory curiosity into a practical technique.

The GE group made several successful flights that winter, proving that dry ice seeding did induce precipitation under certain conditions. But the project soon outgrew them; a full-scale cloud seeding program would require personnel, aircraft, and other resources that the lab did not have. Some far-sighted individuals in GE's legal department also pointed out the modifying the weather could involve the company in a great deal of trouble.

In February 1947 the Army Signal Corps, the Office of Naval Research, and the Air Force came to the rescue with aircraft and support for the seeding efforts. Schaefer dubbed the operation "Project Cirrus" because its goal would be to transform supercooled water droplets into ice crystals, the stuff of cirrus clouds.

Given the goals of the project, it was natural for the researchers to go to New Mexico. Although a substantial part of Project Cirrus, including the official headquarters, remained in the Northeast, the most rewarding results were obtained in New Mexico. Hundreds of clouds in the Albuquerque and Socorro areas were seeded during the course of the project.

The GE researchers, who came to New Mexico in 1948, were lured by the radars, cameras, and expertise of the

Kiale Maynard, a G.E. Labs technician. New Mexico Tech Archives.

Irving Langmuir (l) and Bernard Vonnegut on the desert near Socorro. Photograph courtesy of Charlie Moore.

R&DD, but they were attracted most of all by the unique thunderstorms of the region. During the summer months in central New Mexico, almost every day dawns with a clear blue sky. But by 10 or 11 A.M., puffs of cumulus clouds begin to appear over the mountains. By early afternoon on a good day the solid white mass of a cumulus cloud casts a shadow on the mountaintop. With the right amount of water vapor in the air, the correct degree of solar heating, and a bit of luck, the cumulus cloud continues to develop into a full-fledged thunderstorm. (Ironically, one of the advantages of the area is that the first two conditions are marginal. If there were much more water vapor and heating, the storms would grow too big too quickly for convenient study.)

Thunderstorms can develop in two ways. The giant ones that deliver torrential rains, large hail, and an occasional tornado are usually frontal storms. They arise when two large air masses of different temperatures collide, and they move with the fronts that gave them birth. Under some circumstances, all the storms along a front can turn into a mesoscale convective system, a kind of giant thunderstorm composed of many clouds. These storm systems, only recently recognized for what they are, cause more damage than any weather phenomenon short of the hurricane—which itself is a system of thunderstorms.

The scientists who study these big storms have to mount their equipment on trucks and chase the clouds down the highway. Those conditions are fine for studying the outward aspects of a storm, but they place severe limitations on the examination of its inner workings. Studying frontal storms has been likened to looking into a washing machine during the spin cycle and trying to figure out what articles of clothing are inside.

While the big frontal storms are spectacular and fascinat-

A midmorning cumulus cloud becomes a rainstorm and finally an anvil cloud in central New Mexico. Marvin Wilkening photographs.

ing to study, they have never really been the business of researchers in New Mexico. The scientists there are interested
in orographic air-mass storms, which arise over uplifted terrain because of solar heating. Although these storms get
plenty of respect from the people below, as thunderstorms
go they are rather puny. A typical one might reach 30,000
feet, and is generally the only storm for several miles around.

These clouds are not only small and isolated, but also
relatively stationary. With no large-scale winds to blow
them around, they tend to move only a few miles during
their lifetimes. This made them ideal for Project Cirrus, just
as it makes them ideal for the kind of studies carried out
today at Langmuir Lab.

The convection processes that build the classic New Mexico thunderstorm get their start in midmorning as the sun
heats the local terrain. As the air within a meter or so of the
ground gets warmer, it also becomes thinner and begins
rising. In a process known as orographic lift, the terrain acts
as a guide for the rising air. Parcels of warm air balloon up
the slope until they reach the peak, where they begin to rise
vertically.

The rise of the parcels of warm air creates turbulence in
the surrounding air by displacing the cooler, denser air
above. The cool air has nowhere to go but down. Soon the
warm, humid updrafts are surrounded by downdrafts of
cooler air.

Relative humidity—the ratio of the amount of water
vapor in a parcel to the greatest amount it could hold at that
temperature—then comes into play. When the relative humidity is approximately 100 percent, the parcel is saturated
and can hold no more water vapor. The temperature at
which this occurs is known as the dew point.

As the parcel rises, its constant expansion helps it become
saturated. Since gases cool as they expand, the parcel is con-

stantly approaching the dew point. Non-saturated air cools at about ten degrees Celsius for every kilometer that it rises. Sooner or later the dew point is reached, and the water vapor can no longer remain a gas.

But the excess water vapor is still there, and once saturation occurs it begins to condense onto tiny particles in the atmosphere. These particles, called condensation nuclei, may be solid—dust, for instance—or liquid. When large numbers of the resulting cloud droplets are distributed throughout the parcel, they scatter enough light to become visible. (Even the most ominous-looking cloud is mostly composed of air; the water vapor and ice particles may weigh a million tons, but account for only a few millionths of the cloud's volume. By human standards, a cloud is much more of an optical phenomenon than a material thing.)

Once the air has become saturated, the rate of cooling as it rises becomes slower, for a reason grounded in the basic difference between a gas and a liquid. A gas, such as water vapor, has more thermal energy in it than a liquid. Thus, when the water vapor condenses, the excess energy has to go somewhere; it ends up heating the nearby air. This heating partially compensates for the cooling caused by the expansion. The parcel of air, still rising, is now cooling at only six degrees per kilometer, a figure typical of New Mexico clouds.

That figure, known as the moist adiabatic lapse rate, is the key to the growth of a cloud. The parcel of air, now far from the original heat source on the sunny ground, is cooling off. But the dry air through which it is moving is still getting colder at the dry adiabatic lapse rate of ten degrees per kilometer. The net result is that the cloud is still warmer and less dense than the surrounding air, so it continues to rise.

The rise and fall of neighboring air parcels causes turbulence, so that the warm, moist air and the cool, dry air mix, lowering the amount of water vapor per cubic meter. In addition, rising parcels sometimes rise too high, whereupon they sink and their water droplets evaporate. These two factors can combine to make the morning's first clouds literally disappear into thin air.

Sometimes, though, a convective system of rising air grows large enough to withstand a little mixing and evaporation. If there is enough moisture in the air, a fair-weather cumulus cloud forms. Cumulus clouds, the familiar cauliflower clouds of summer, look simple and peaceful from the ground, but a flight around one reveals billowing white masses boiling upwards on a scale that dwarfs the airplane. Each cumulus cloud is made up of many parcels of air, some on the rise, others on the way down. This continuous rise and fall of air parcels gives the cloud its characteristic billowing top. The cloudbase remains flat, marking the height at which condensation first takes place.

Even as scientists crane their necks hopefully at the cloud, many things can stunt its growth. A severe wind shear, a radical difference in direction between strong winds aloft, can limit the cloud's growth. Even worse is a temperature inversion, which means a stable air layer that the cloud turrets may not be able to poke through. And there simply may not be enough heat and moisture involved to support a big storm.

But if the conditions are right, one of these cumulus clouds, peaceful, rainless, and unelectrified, keeps on developing. Bigger and more vigorous than its neighbors, able to control the convective scene over many square miles, it grows in size and complexity: a cumulus congestus, dominating its region of the sky.

Within the cumulus congestus, several processes are happening at once. The cloud droplets collide with one another. Some of them bounce away, but others carry minute, opposing electric charges that help them coalesce, forming bigger droplets. As these droplets grow, some of them become involved in a tug of war between the updrafts and the force of gravity. As the droplets collide and coalesce, they become drizzle drops, and gravity begins to win the contest. The big drops fall, picking up any smaller cloud particles they happen to run into. Radars begin to show a precipitation echo within the cloud as rain begins to fall.

If the cloud keeps growing, it can develop into a cumulonimbus, a big, highly convective thundercloud. At this point, it becomes electrified, a process that still causes scientific controversy. But Langmuir and company were not primarily interested in thunderclouds in the late forties. They wanted clouds for rainmaking.

The Project Cirrus team arrived in New Mexico in October 1948, having been warned that the rainy season was already over. A popular local aphorism holds that anyone who tries to predict the weather in New Mexico is either a fool or a newcomer. Their arrival was greeted by rains. Torrential rains, causing flash floods down the arroyos. The anomalous weather was a bit embarassing for the R&DD staff, but it certainly demonstrated the area's potential as a site for atmospheric research.

Although the rain was a good omen, it wasn't a necessary condition for Project Cirrus. Cloud seeding works by causing minute ice crystals to form inside a supercooled cloud; it does not matter whether it is warm enough below for the falling crystals to melt.

Journalists in parched New Mexico were enthused about Project Cirrus. They quoted Mark Twain, as reporters are apt to do, then triumphantly refuted him. Finally somebody had done something about the weather! Some of the people who wrote about cloud seeding were well informed and cared about the scientific accuracy of their reportage. Bal-

An anvil cloud over the Magdalenas, August 14, 1962. Charles
Treseder photograph.

ancing those conscientious individuals was the Albuquerque journalist who wrote that Langmuir had gone up in a B-17 and made a cloud disappear and that area residents should not be alarmed at the sight of oddly-shaped clouds.

Workman and Reynolds found themselves in the midst of all this weather modification activity, they were distracted from their thunderstorm and lightning research. There remained a research effort called the Thunderstorm Project, complete with a Thunderstorm Laboratory at the School of Mines, but the days of sitting in a field and watching a distant thunderstorm were far from their minds.

One day, Reynolds was in a B-17, flying around the mountains near Albuquerque in search of an appropriate cloud. "Drop the ice," he ordered when one was found.

They dropped the ice—all in one solid 50-pound chunk. The experiment involved water ice, not dry ice, and the shaved material had frozen together in the flight through the high, cold air. Fortunately, they were over an unpopulated area in the high desert southeast of Albuquerque. But they were also near Kirtland Air Force Base, and someone had neglected to inform them about the bomber. Unbeknownst to Reynolds, a fighter had been scrambled to inspect the mysterious aircraft. The fighter was flying behind and below the B-17 when the ice was dropped. For once in aviation history, Murphy's Law did not apply. Some red faces were in evidence, but both the fighter and the careers of the B-17 crew escaped damage.

Despite such aerial misadventures, the cloud-seeding project was far more than a series of "let's try this and see what happens" experiments. As Project Cirrus grew, so did its scientific sophistication. Specialized apparatus was developed, including an automatic cloud-type identifier and an array of devices to control the sizes of the seeding particles. Planes laden with instruments flew in and around the

Workman's mobile silver iodide smoker in operation. New Mexico Tech Archives.

A Project Cirrus B-17 bomber with some optional equipment.
New Mexico Tech Archives.

seeded clouds so that scientists could tell exactly where the seeding materials were going and what effect they were having.

The idea of making rain, however, was certainly not the invention of Langmuir and Schaefer; what stereotypes the American Indian if not the rain dance? The celebrated rain-makers of the American West went the Indians one better, firing cannons and skyrockets to give drought-stricken sod-busters a good show for their money. During World War II, a Luftwaffe pilot even dropped 300 pounds of sand into a cloud over Yugoslavia, perhaps in a misguided attempt to provide condensation nuclei. But the days of shamans, confidence men, and earnest dilettantes had given way to the day of the scientist. Never mind whether Project Cirrus was going to help bring rain to New Mexico, the scientists were sure that the knowledge to be gained in the attempt would be a good return on the investment.

They experimented with silver iodide as well as dry ice. That was a more peaceful pursuit than dry-ice seeding, because silver iodide could be vaporized on the ground and borne up by the updrafts of a promising thundercloud. To mobilize the research, a 1948 Oldsmobile coupe was equipped with a flamethrower-like silver iodide smoker.

Seeding from ground level was nothing new; on an expedition to a Navajo village, Workman and Reynolds found that the Indians had been doing it since time immemorial. First, the scientists demonstrated nucleation with a portable cold box. Unimpressed, the Indians responded by burning the charred bark of a tree that had been hit by lightning. The scientists were using applied physics, while the Indians were using sympathetic magic, hoping that like would bring like. Both have claimed success.

The silver iodide seeding efforts eventually sparked a major disagreement between Langmuir and some of his colleagues. Langmuir, who was back in Schenectady, New York, at the project headquarters, had some R&DD people do a silver iodide experiment for him in Socorro. Beginning in late 1949, he had them seed the local clouds eight hours a day, three days a week. Langmuir eventually noticed a pattern. Several days after clouds were seeded in Socorro, it would rain in the Ohio Valley.

Come Thanksgiving 1951, the R&DD employees asked for a few days off, and Reynolds agreed. Langmuir was furious. Where was their dedication? But his temper cooled down when he checked the weather reports. Sure enough, a few days after they stopped seeding in Socorro, the expected rainstorm in the Ohio Valley failed to occur.

Langmuir came to the conclusion that his seeding project in Socorro was causing the Ohio Valley rains. It was a daring step, because researchers learn early that correlation alone does not imply causation. But to Langmuir, the pattern was stretching the bounds of coincidence.

Whether true or not, it was a disturbing conclusion. If local activity in the New Mexico desert was causing rain by the banks of the Ohio, something was going on that could not be explained. Langmuir arrived at the tentative conclusion that some type of resonance in the atmosphere was at work, that the seeding in Socorro was doing something to the atmosphere rather like striking a tuning fork. Such long-period resonances were known to exist, but little else was known about them.

Langmuir finally decided to present his results at a conference. His colleagues tried to dissuade him, reminding him of the local nature of the research effort and the great distance between Socorro and Ohio. Finally they simply told him that he was talking nonsense. He presented his results anyway.

No one laughed out loud at the idea—after all, Langmuir was the patriarch of the entire field, and the weather pattern existed—but the break between Langmuir and the R&DD

Inside and (*right*) outside an instrument trailer, circa 1957. New Mexico Tech Archives.

scientists had been made. By the late 1950s, cloud seeding
had gone completely out of favor in New Mexico. Workman
had become a nonbeliever, and to a certain extent so had
Reynolds. But Langmuir remained a believer until his death
in 1957.

Does cloud seeding work? There is no doubt that introducing dry ice or silver iodide into a supercooled cloud in
the proper way causes widespread nucleation and ice crystal
formation. Plenty of hard physical evidence verifies that
phenomenon. There is also good reason to suspect that
Project Cirrus brought down hundreds of millions of tons of
rain in New Mexico that would not have fallen naturally.
The spreading disenchantment with rainmaking was caused
not by doubts about its effectiveness so much as by questions about its predictability and ethicality.

A 1978 case involving the Adolph Coors Company, a
Western beermaker, provides a sterling example. The farm-
ers around Alamosa, Colorado, where barley was raised for
Coors and other customers, demanded and received a hearing before a special master. They claimed that the company
was deliberately overseeding the clouds to prevent rain from
falling on Coors barley at harvesttime. This, of course, also
kept it from raining on neighborhood farms. Coors claimed
that they were merely trying to suppress hail. Charles Moore,
professor of atmospheric physics at New Mexico Tech, was
called in to testify. He testified that the hail-suppression claim
was indefensible, and that overseeding a cloud to the extent
required by Coors could indeed cause the formation of too
many ice crystals too small to reach the ground, thus sup-
pressing rainfall. This view was accepted by the special master, who ruled against the cloud-seeding operation.

The special master's view was apparently also accepted by
the Soviet Union. There have been reports of cloud seeding upwind of the Chernobyl reactor site during cleanup operations there. The goal, evidently, was to suppress rainfall

that might wash radioactive contaminants into the soil and streams.

In October 1947, Langmuir himself was involved in an episode much bigger and more controversial than the Coors case. He had come up with the idea that scientists might be able to defuse a hurricane by overseeding it and thus disrupting the dynamics of the individual thunderclouds in the storm. Out went his team to penetrate and overseed a hurricane that was off the Florida coast and heading for Bermuda.

Something may have happened; that was the only firm conclusion that could be drawn. The newly seeded hurricane, which had been spending its fury on the empty Atlantic Ocean northeast of Jacksonville, quickly gained strength, made a 90-degree left turn, and roared ashore near Savannah, Georgia. A Miami journalist described the seeding as "a low Yankee trick." Langmuir was thoroughly pleased with himself, but GE's long-ulcerating lawyers were not pleased with Langmuir. The company immediately told their Nobel laureate not to boast about his achievement before the statute of limitations ran out.

But what did the low Yankee trick really accomplish? No one knows, and perhaps no one will ever know. A hurricane, essentially a vicious circle of thunderstorms moving as a collective entity, is one of the most freakish of natural phenomena. Perhaps the seeding gave the hurricane its new strength and steered it back towards land. Then again, perhaps the storm was going to make a 90-degree left turn anyway (hurricanes have performed maneuvers much more radical than that) and the seeding kept it from being even more powerful than it was. And perhaps the seeding had no effect on its course at all.

In any case, cloud seeding was bound to result in a great deal of disappointment, simply because public expectations were so high. A 1949 newspaper piece entitled "R&DD Technicians Saddle Thunderstorms" read, "Perhaps some day the weather man will be able to literally order the weather for the day."

Both scientists and the public have learned since then. One of the best markets for the latest and fastest supercomputers is in numerical modeling of the weather. The variables involved are forbiddingly numerous, and no one really knows where to draw the line. The data pour in constantly from satellites, from ships at sea, from ground stations the world over. How much is useful information and how much is noise? Does the course of a butterfly's flight in Peking affect the weather in Hoboken a week later? Some scientists have suggested that it might. Today only the most sensational of the tabloids at the checkout stand would dare to predict the custom-ordering of the day's weather. How could we control the weather without first understanding it fully— and more importantly, how would we dare? The hubris of the Fifties has given way in the Eighties to chaos theory and to numerical modeling.

But, as Reynolds recalls, those were happy days to be a scientist—the days of the postwar technological miracle and the autogyro in every garage. All eyes were on the glory of the scientific future in a jet-propelled America that had split the very atom. Surely scientists could learn to control the weather!

Reynolds, who left Tech in 1955, witnessed Langmuir's tenure as a true believer in rainmaking, and he watched as Workman's enchantment flowered and died. For the past thirty years, he has been the State Engineer of New Mexico, the official in charge of water resources: a man of strong political influence who is known to look unfavorably upon tampering with the clouds. Nobody does much cloud seeding in New Mexico anymore.

4

The Plains Of San Agustin

Although E. J. Workman neglected thunderstorm research during his fling with rainmaking, a few diehard atmospheric electricians at the School of Mines continued to probe the inner workings of the thunderstorm. Among them were Steve Reynolds and, later, one of Workman's former students, Marx Brook.

Today, Brook could be called an elder statesman of atmospheric physics, but that honorific belies his tendency to carry on verbal battles with his colleagues through the halls and into the elevator—in a Brooklyn accent that leaves no irony unturned, and often at paint-flaking volume.

Since 1978 Brook has been the director of the R&DD, a senior administrator with little time for active research. But he remains a professor of physics and a mentor to graduate students, reigning over the R&DD as a benevolent tyrant. The spirit of Workman still walks the halls of New Mexico Tech.

In 1954, when Brook joined the Thunderstorm Project, Workman was alive and well and preoccupied with introducing foreign substances into cold clouds. Brook was no stranger to the work, having earned his bachelor's degree at UNM under Workman and Holzer during World War II. After graduation, he had stayed on until 1946, working both on charge separation in thunderclouds and on proximity fuses.

After the war came a four-year sojourn at the University of California at Los Angeles, where he earned his master's and doctoral degrees in physics. At UCLA he studied various properties of nitrogen gas and published three papers on the subject in the *Physical Review*. Soon, he found his interests turning back toward lightning research, so he asked his old professor for a job.

Brook came to New Mexico Tech with the idea that he was going to be strictly a researcher. But such aspirations are more commonly stated than fulfilled, especially at small colleges; within a year, Workman asked him to teach a few classes. Brook soon found himself to be an assistant professor of physics and has been a member of the teaching faculty ever since.

With senior researchers busy wafting silver iodide into the bottoms of clouds and dropping dry ice onto their tops, Brook found an opportunity. It lay on the Plains of San Agustin, an ancient dry lakebed west of the Magdalena Mountains. Today, the Plains of San Agustin are an important center of research, for the National Radio Astronomy Observatory's Very Large Array radiotelescope is located there. But Brook and a few other physicists from Socorro beat the astronomers there by thirty years.

In 1952, Reynolds and a colleague had set up a small network of "field mills" on the plains. Although the term encourages thoughts of windmills and grindstones, field mills are actually small rotary instruments—think of a coffee can with a small fan inside—that measure the electric field caused by charged clouds. By burying the field mills so their business ends were flush with the ground, the researchers could gather data on the charging of the clouds overhead. Using several field mills allowed them to figure out, by triangulation, how the field strength varied with position and distance from the storm. It was not an exact determination. A single field mill cannot tell the difference between a charged thundercloud a kilometer overhead and a technician rubbing a piece of Teflon a meter away. But two field mills do a fair job, and several of them spread out over a wide area give a good approximation.

Also on the plains were time-lapse cameras that could record the progress of the storms. But one of the most promising new tools the scientists used was radar. The wartime need to detect enemy planes had provided a windfall for atmospheric physics. At last scientists could see through clouds, picking up rain shafts, hail, layers of ice crystals, and other things that had to be located and measured.

The project was soon removed from the plains. Working there had not been without its hazards. Brook once found a rattlesnake curled up inside an instrument package. Reynolds went into a bar in the town of Datil, made an innocent remark to a rancher about the progress of a private rain-making project, and got punched in the nose. But the real reason for the move was that the plains were not quite the right place for all this equipment and excitement; storms were not frequent enough and were too likely to travel.

By the time Brook arrived, the work was shifting to Mount Withington, at the southeast end of the plains, where small local storms occurred almost daily during the rainy season. One of Brook's early tasks was to help set up a radar on the mountain in an attempt to correlate the first radar echo in a storm with the first detectable electric field beneath it. A good correlation would be a rather broad hint—although not an ironclad proof—that whatever reflected the radar was also generating or gathering charge.

Obtaining a correlation between a radar echo and initial electrification took quite a while. Gathering data in atmospheric physics always does. The electronic equipment often has to be built by the scientists themselves; it is notoriously temperamental, and learning to use it efficiently takes time. Nature is, if not deliberately uncooperative, at least

Steve Reynolds (in cowboy hat) and E. J. Workman (in checkered shirt) prepare a balloon-borne instrument package for launch.

monumentally indifferent. A bone-dry week will end with a flash-flood warning, generally on the researchers' day off. There is no instrument an insect cannot crawl into and no cable a determined rat cannot gnaw through. Rarely will a field operation yield useful data before the third year; sometimes four or five years of frustration are in order.

Brook and Reynolds were lucky. They had results to present at a conference in 1956. Yes, the radar showed something in there at the same time that we recorded a foul-weather electric field on the ground. No, we didn't fly through the cloud and investigate firsthand, although we would have liked to. Of course, we need more money to continue this research. Back they went to pursue their work further.

By the time they returned, another pair of researchers had become active on Mount Withington. Bernard Vonnegut had left GE Labs to join the Arthur D. Little Company, a scientific thinktank in Cambridge, Massachusetts. There he met Charles Moore, another scientist interested in atmospheric research.

Charlie Moore was among the last of a dying breed. Today, the Ph. D. is a researcher's basic union card. Moore had only a bachelor's degree in chemical engineering from the Georgia Institute of Technology in Atlanta. His qualifications for investigating the fundamental theories of atmospheric electricity were largely nonacademic. Armed with neither dissertation nor multiple sheepskins, he made do with peripatetic curiosity, an astonishing amount of energy, and the ability to keep smiling through the barrage of chemical-engineer jokes that flew about him every time he picked up a pipe wrench.

Moore also had a singular past. He had done graduate-level work in the Army Air Corps during World War II en route to becoming chief weather equipment officer for the China-Burma-India theater. After the war he went to work for General Mills, which had branched out from the cereal business into the balloon business. Moore was the pilot in the first flight test of a manned plastic balloon. He took the balloon up to 10,000 feet, attached to it by his parachute harness. He had several patents to his name and was one of the co-discoverers of traces of water vapor in the atmosphere of Venus—a discovery made from a balloon gondola.

Vonnegut and Moore made something of an odd couple. Moore is garrulous and energetic, with high-voltage blue eyes and a nonstop train of thought. Vonnegut is given to long thoughts and thousand-yard stares. But they had scientific problems in common, and soon they shared a field project. They first went to Mount Withington in the summer of 1956.

They went there in part because Vonnegut had made himself one of the most celebrated heretics in atmospheric physics. Vonnegut had a new and highly unpopular hypothesis of cloud electrification, and his investigations of it had outgrown the laboratory. A whole cloud was needed, and he knew from experience that Mount Withington was the place to find one. Moore came along to add his meteorological expertise to Vonnegut's knowledge of physics.

Many suggestions have been made to explain how a cloud might become electrified. Most of them have been shown to be wrong for various reasons. Any new hypothesis has to account for the fact that thunderstorms are almost always negatively charged on the bottom and positively charged on top. The electrification scheme presented by a new hypothesis has to involve physically plausible processes and must cause potential differences large enough to result in lightning. A good hypothesis should also be useful in explaining how thunderstorms help maintain the electrical budget of the earth, which is negatively charged and has been for as far back as anyone can tell.

The question of how a quiet cumulus cloud becomes a giant electrical battery has perplexed scientists for a century or more. Lord Kelvin, the great English physicist, decided in 1860 that falling rain somehow charges the cloud. Although Kelvin had no real data behind his hypothesis, it was plausible and carried credence because of his reputation as one of the scientific giants of the century. His hypothesis was handed down through the years, occasionally refined and expanded, until it became one of the basic assumptions from which atmospheric physicists worked.

But Kelvin's hypothesis merely explained how electric charges of different polarities could become separated within a cloud. It gave no clue to where the charge came from in the first place. That was explained by various other means. One of the more popular explanations was the Workman-Reynolds effect. Workman and Reynolds had found that a potential difference existed at the border between an ice crystal and the water that was freezing onto it. They proposed that when water and ice met in a cloud, some of the water, with its negative charge, spattered off, leaving the positively charged ice behind.

It was later found that, although the Workman-Reynolds effect is real, the negative charges are bonded to the interface between the ice and the water and do not spatter off. Lightning has also been observed in "warm" clouds—ones that do not contain ice crystals—as well as in volcanic dust clouds. Nonetheless, many scientists still believe that the effect could account for charge separation in a typical storm cloud, although no single mechanism is likely to account for all charge separation in all clouds. The Workman-Reynolds effect is often combined with Lord Kelvin's hypothesis: falling rain carries the negative charges to the base of a cloud, leaving the positive ones at the top.

Vonnegut agreed that charged precipitation plays a sig-

The Vonnegut and Moore mobile laboratory at Langmuir Lab site, 1965. New Mexico Tech Archives.

nificant role in cloud electrification, but he believed there was more to it—that charge could also be brought into the cloud from outside. He proposed in 1953 that positive space charge—clouds of positively charged particles known to exist above the ground—was carried to the top of convective clouds by the updrafts in the middle. The next step, as he saw it, was that the downdrafts around the clouds also carried negative charge downward.

Vonnegut's theory was ridiculed. In the mid-50s, many atmospheric scientists believed that there were no downdrafts during the development of a storm and that the winds converged instead of diverging at the cloudtop. If confirmed, this would have meant the end of the convective hypothesis. Vonnegut and Moore later found that a French scientist, Gaston Grenet, had proposed a somewhat similar mechanism in 1947. Grenet's theory had been ignored and was almost forgotten by the time Vonnegut learned of it in 1957.

Vonnegut had already created a stir by proposing that strong updrafts into the cloudbase and divergent winds at the cloud top even existed in developing thunderstorms. By seriously proposing a hypothesis that most of their colleagues ridiculed, he and Moore left themselves no way out. They had to admit to a serious error or do whatever was necessary to vindicate their viewpoint. A few years later, a trip to Withington gave Moore a splendid, if serendipitous, opportunity to refine their hypothesis.

One drizzly afternoon in the summer of '58, the cloudbase enveloped the top of the mountain as Moore packed up his instruments to return to Boston. It had been a quiet cloud electrically—but suddenly lightning struck a nearby tree.

The event came as quite a surprise. Even more surprising was what happened a minute or two later. The instrument trailer Moore was in, having nearly been struck by lightning, was drenched by a gush of brief, heavy, localized rainfall. As he watched the storm's late-afternoon convective tantrum, lightning struck six more times in the vicinity. Each stroke was followed by a gush of rain.

Now his curiosity had been piqued. Obviously there was some connection between the lightning and the rain. This fact could be explained in terms of conventional theory: the falling rain, caused by God-knows-what, had brought negative charge close enough to earth for lightning to occur. But the more he and Vonnegut thought about it, the more the time lag seemed to suggest that the lightning might be causing the rain gushes. Very well and good, but they still had to prove it. Both rain and lightning were known to originate a kilometer or more up in the cloud. Moore had observed that the lightning came out of the cloud before the rain, but once again, correlation did not imply causation. Lightning travels much faster than raindrops. Perhaps the rain caused the lightning, just as orthodox thinking indicated, and the lightning simply beat the rain to the ground.

A long tradition existed in atmospheric physics of trying to relate rain gushes to lightning. Kelvin's hypothesis explained how falling rain could cause lightning. Earlier, Robert Hooke had noted a relationship between rain gushes and lightning in 1664. The Roman Lucretius had observed the correlation in 58 B.C. and concluded that thunder caused rainfall. Moore, ever on the alert for new sources of data, claims to have found a reference to the phenomenon in some translations of the Book of Exodus. In keeping with this tradition, Vonnegut and Moore figured out how lightning might trigger rain gushes, then published their own hypothesis.

Their explanation rests on the way that they envision the origin of a lightning channel high within a cloud. They be-

lieve that the channels do not branch so finely as to reach
every charged droplet. Rather, they suggest that the top of a
lightning stroke is only broadly tree-like, with branches but
no twigs. The branches are channels densely packed with
positive charge. They reach into cloudmass that contains
negative charge, but at a comparatively low density.

Because oppositely charged particles attract each other, the
positive particles move into the cloud and combine with negative ones. The bigger particle that results still has a positive
charge, but not as much as before. These bigger, positively
charged particles are attracted further into the cloudmass,
where they hit and coalesce with more small ones.

As the process repeats itself, the particles keep growing.
Before long they have grown into droplets and have been neu-
tralized almost completely. As the droplets fall, they keep
merging with whatever is in the way, but electrical charge no
longer has much to do with it; all that is left is a mechanical
process of collision and coalescence. Faster and faster these
big drops fall until they come down as a rain gush.

Or so said Vonnegut and Moore. They worked out the
mathematics of this process, which was quite plausible in
theory. But they needed more than theory to prove their
case to the scientific world. What they needed was an instrument that could tell them the precise order of events inside
the cloud. The radars of that day were simply not up to
the job.

In the meantime, some of the opposition to Vonnegut and
Moore's theory was coming from Tech. Brook, an adherent
to the conventional variation on Kelvin's hypothesis, does
not doubt that rain gushes can occur after lightning. But he
has also reported on a "lightning gush" in which lightning
occurred after a heavy burst of hail. Why aren't rain gushes
observed after lightning more often? Brook asks. He consid-
ers the phenomenon to be fairly uncommon; such organiza-

tions as the National Severe Storm Center in Norman, Oklahoma have searched for its origins in vain while studying other aspects of the relationship between rain and electrification. Brook, who uses phrases like "tilting at windmills" and "beating a dead horse" to describe Moore's outspoken advocacy of the new theory, sometimes even hints that Moore enjoys his self-perceived martyrdom. In any case, Vonnegut and Moore have been tilting for a quarter of a century, and so far the windmills are still winning.

Another project that would have to wait was a direct test of Vonnegut's convective hypothesis. The controversial hypothesis suggested a good experiment: release space charge below a convective cloud and see what happens. A variety of schemes for his would be tried over the years, yielding suggestive partial results but no concrete evidence.

In the meantime, Moore and Vonnegut found other things to do. Moore tried ballooning on Mount Withington, with the idea of flying straight up into a thunderstorm in the gondola of a helium-filled plastic balloon. He was quoted as saying, "If the electric field gets too dangerous, we'll toss ballast and go through the top of the cloud." That turned out to be wishful thinking; he and his co-pilot, Lieutenant Commander Malcolm Ross, ended up miles from Withington, having proven once and for all that a balloon is not an appropriate manned research craft for thunderstorm work.

After three years of commuting between Massachusetts and Mount Withington, Vonnegut and Moore had some intriguing and significant results. Their principal work on Withington had been an extension of Reynolds's and Brook's studies of the relationship between precipitation development and organized electrification in growing clouds. With instrumented captive balloons flown into the clouds and a radar directly below, Vonnegut and Moore found that organized electrification could be detected before the raindrops

Charlie Moore (left) presiding over the removal of the Tech hot-air balloon from a barbed-wire fence. New Mexico Tech Archives.

had grown large enough to be responsible for it. They also found again that gushes of rain often fell soon after nearby lightning.

Other interesting phenomena were discovered: the end-of-storm oscillation (a lazy foul-to-fair-to-foul alternation of a dying storm's electric field) and curious electric-field excursions just before the arrival of rain. But despite their discoveries, the two were still not much closer to a thorough and basic understanding of the thunderstorm electrification process.

As Vonnegut and Moore surveyed the work that had yet to be done, they were dismayed by the prospect of commuting from Cambridge every year and working out of trucks all summer. September 1958 found Vonnegut at a cloud physics conference in Australia and Moore atop Withington packing up their equipment between rain gushes. With the truck loaded, Moore and Brook sat in a bar in the nearby village of Magdalena. Moore proposed that they jointly build a small, permanent lab on Withington. Brook was interested in the idea and soon passed it on to Workman, who, unknown to Moore, had once proposed such a laboratory on Sandia Crest near Albuquerque.

While all the scientists agreed that a permanent lab would be a good idea, they could not agree on where to put it. The New Mexico Tech scientists wanted it to be in the Magdalena Mountains, which were just east of Mt. Withington and the San Mateo range. This location would greatly simplify the logistics of supporting the lab and would permit line-of-sight microwave data transmission to Tech. Moore preferred Withington, where summer thunderstorms occurred three times more often than in the Magdalenas.

For a time, two independent efforts were made to build a mountaintop thunderstorm lab in the area: Moore's and Workman's. In mid-September, Workman wrote to the chief of the U.S. Forest Service to propose a joint New Mexico Tech-Forest Service project to build a road to the top of South Baldy Peak, the highest point in the Magdalenas. The Forest Service administrators in Washington were less than enchanted with the idea, but Workman kept trying.

In the meantime, Moore designed a lab building for Withington and secured the cooperation of forest rangers there. But when he went to the National Science Foundation for money in 1960, he ran into a stone wall. The NSF, which is supposed to fund university-affiliated research only, had no intentions of supporting such an effort on behalf of the Arthur D. Little Company. They liked the idea, however, and encouraged Moore to leave the company and become associated with Tech.

That looked like a difficult business for Moore, who had no Ph.D. and was a controversial figure to boot. Moore and Vonnegut gave their support to Workman's project, despite South Baldy's marked inferiority as a thunderstorm producer. As it turned out, however, Moore would have both a permanent mountaintop laboratory and a tenure-track associate professorship at New Mexico Tech by 1965.

5

The Research Finds A Home

So it was decided, with Moore's somewhat reluctant blessing, that the mountaintop laboratory in New Mexico Tech's future would be on top of South Baldy. Only two problems remained in the early-planning phase: obtaining permission from the Forest Service to blast and bulldoze a road through a national forest in order to erect a four-story building, and getting the money for the road.

Workman approached the problems with typical cunning and ingenuity. The mountain was in a national forest, was it not? So who should build the road? Why, the Forest Service, of course!

When Workman broached the idea in 1958, Regional Forester Fred Kennedy, like his superiors on the Potomac, did not favor the idea. He said that the administrative needs of the Forest Service would not be particularly well served by such a road. Kennedy did state, however, that since South Baldy was occasionally used as a forest-fire lookout, it would be nice to have a road. Finally, he promised that a ranger would be in touch later.

Workman had enough experience with bureaucracy to know when he was being given the brush-off, and enough patience to wait. In July 1959 he gave up on Fred Kennedy and wrote again to the chief of the Forest Service. Workman and the chief were acquainted, a fact Workman stressed in the letter. The reply, though, came from the assistant chief, who stated that most such roads were built for access to timber and that lack of funds limited the number of roads that could be built.

The reply was plausible, but Workman realized the real problem was one of jurisdiction. The Forest Service is officially responsible for a number of things, but supporting physics research is not one of them, so it wanted nothing to do with the lab. (This attitude would haunt the place for years.)

Frustrated in his attempt to get results from Washington, Workman went back to Fred Kennedy. This time, Kennedy promised to send a survey party to Baldy as soon as possible to estimate the cost of the proposed road.

In April 1960, Kennedy finally reported that several reconnaissance studies had been made in the area and that a review of the route, along with a preliminary cost estimate, were being prepared. By then, Workman was getting thoroughly tired of the delays, so he set about pulling every official string within reach. Soon the Chief of Naval Research and the head of the Arthur D. Little Company were writing to the Forest Service about the need for such a laboratory. The Forest Service replied that the ten miles or so of road could cost more than $100,000, and that Workman's need for the road was not pressing enough to squeeze out that kind of money.

As seen from Socorro rather than Washington, building a road up Baldy did not look like a $100,000 proposition. Workman's reply was couched in the courtly language he always used in dealing with high officials, but the point was perfectly clear. He said, in effect, "You guys are full of beans. I'm checking this out myself." And so he did. While Tech crews hiked up and down South Baldy, Workman wrote to New Mexico's senior U.S. senator, Dennis Chavez, telling him how much New Mexico's universities were contributing to defense research. He slipped in a mention of the problems with the Forest Service. It was not a plea for help—just a reference to the problem. Senator Chavez replied that funds should be available soon.

But not even a senator could light enough of a fire under the Forest Service to please Workman. By March 1961, he had a firm commitment to build a laboratory, but none to fund a road to it. Under this peculiarly intense duress, he gave up on the Forest Service and asked for $35,000 for the road in the final proposal to build the lab itself.

By making an estimate only one-third as large as the official one from Washington, Workman had thrown down a considerable challenge to himself. He sent physics professor Marvin Wilkening, mining professor George Griswold, and other faculty members up the east side of Baldy once again. This time they spent weeks hiking back and forth over the rugged terrain. The professors soon came to appreciate the irony of their mission. At the same time that they were enjoying the fresh air and beautiful scenery, they were trying to figure out how to blast and bulldoze a road suitable for diesel trucks all the way to the top of the mountain. After they examined the terrain and made a rough plan, Lamar Kempton of the R&DD would bring in surveying and construction teams to finish the job.

But Kempton's crews could not go right in with chainsaws and dynamite. Conservationists, hunters, and ranchers all voiced their opposition to the plan at a public hearing on August 6, 1961. This time, though, the senatorially-prodded Forest Service was on the side of the scientists. Clay Withrow, head of the Magdalena Ranger Station, came out in favor of the road, noting that it would open the area to hunters, hikers, and firefighters. (The local hunters actually did not care for that idea; they did not want to make access any easier for hunters from other areas.) Withrow even offered to attend the next protest meeting to explain the purpose of the road and its benefits to the public. The persuasiveness of Wilkening and Withrow was such that the next protest meeting never took place.

Some of the objections were based on the proposed location of the road. The most beautiful scenery in that part of the Magdalenas is on the east face, an area of sheer cliffs and tall pines along the sides of Water Canyon. Building the road up the east side would leave a ten-mile scar zigzagging up the side of the mountain. Circling around the north end of the mountains past the village of Magdalena and going up the west side of South Baldy would have been technically easier and esthetically less damaging. It would also have made access easier for the Forest Service, which had its ranger station in Magdalena. But it would have added about twenty miles to the researcher's daily trip up the mountain. Practicality won out, and the survey crews went to work.

It was then that Workman unveiled his secret for doing the $100,000 job for $35,000. There was no need to hire expensive outside contractors; Kempton could do the entire job with a crew of experts from TERA, the Terminal Effects Research and Analysis group that was part of the R&DD. Kempton and Kenny Sorensen, his right-hand man from TERA, walked and rode horseback up and down the route suggested by Wilkening and his group. Trying to match the road with the terrain, they decorated the mountainside with dozens of rolls of surveyor's tape, using every available color and finally resorting to combinations of colors.

Finally, one summer day in 1961, Kempton took a forest ranger on a hike up the only acceptable route. The Forest Service agreed to the plan on the condition that Kempton himself would be in charge of everything. The final approval for construction would have to be obtained half a mile at a time, and would depend on the quality of the work done on

Looking down south Water Canyon to the Rio Grande Valley.
Marvin Wilkening photograph.

the previous half mile. Under those terms, the Forest Service raised its official estimate to $220,000, with a projected completion date of December 31, 1967.

The Forest Service should have known Workman better by then. Given enough money, enough manpower, and a free hand to administer both as he saw fit, Workman could probably have finished the interstate highway system by December 31, 1967, and then pitched in on the space program. By February 1962, the road had been roughed out all the way to the top.

Rough was exactly the word for it. Most of the road was cut with bulldozers (one on loan from TERA and two brought in from outside, along with their operators). Everyone had standing orders to make the most of all available daylight; this meant that the pair of hired "cat skinners" lived five days a week in a trailer at the site, while the TERA workers had the dubious privilege of commuting to the site by daybreak. About 5 percent of the road had to be blasted out of solid rock. Kempton did the "powder monkey" work himself.

Along with the road was built the first of the Langmuir traditions: the assignment of nicknames and folklore to inanimate objects. The first switchback was dubbed Lone Pine Turn. Kempton informed the crew that the lone pine at the curve was holding up the sky and that anyone who took an axe to it would be extremely sorry. Above it came Barber Pole Turn, with a tree wrapped in two colors of surveyor's tape, and then Kemtown, named after the boss, where there was room for the Cat skinners to park their trailer. Near the top, the road made an S-curve through a stand of tall pines. The fellow who had to cut the trees observed, "Sure woody up in here." The curve became known as "Sure-Woody" (to the old-timers who remember the origin of the pun) and as "Sherwood Forest."

While the road work continued, Workman applied for Forest Service permission to build the laboratory. Along with the request for a permit, he sent a seven-page supporting statement, with pictures, detailing how the lab would be built and what it would look like. He had designed the structure himself, acting as both architect and engineer. The plan called for four enclosed floors and three decks. The foundation was to rest on bedrock. Workman wanted a building that could shrug off 150-mph winds, three-foot ice accumulations, and as many direct hits by lightning as the mountain could attract.

In March, Workman wrote to Langmuir's widow:

I need to have your permission for doing something that I have wanted to do for a long time. We are building a building under the auspices of the NSF. The building will be located on South Baldy Peak in the Magdalena Mountains. That is a nice sharp peak to the west that was visible from your apartment. The building is going to be a nicely conceived affair, artistically developed and dedicated to the study of atmospheric physics, which subject was brought from very early stages of childhood through adolescence by Irving. All of us want to call this building the Irving Langmuir Laboratory of Atmospheric Physics.

That same spring, Workman rode to the top of South Baldy Peak with Kempton and Sorensen. The day was cold and windy, as springtime in the mountains of New Mexico is apt to be. They sat on the peak to eat their lunch and admire the view. To his surprise and disappointment, Workman could not see Tech from the summit. The top of Socorro Peak, on the western edge of the Tech campus, was in the way. He began to grouse about how much trouble it would be to blow away the top 150 feet of Socorro Peak to permit line-of-sight microwave transmission between the lab and Tech.

A bulldozed portion of the lab road. Marvin Wilkening
photograph.

Kempton had no desire to blow the top off Socorro Peak, cut a notch in it, or do anything else along those lines, so he suggested riding out along the ridge that ran southeast from Baldy. Workman was sure that the knoll on the end of the ridge would be too low, but Kempton insisted that they at least look at it. When they got there, trees sheltered them from the wind, the birds were singing, and they could see straight to Tech. Workman decided then and there to move the lab site.

Workman made some changes in the plans while the lab was under construction. The biggest change was the substitution of one floor and an enclosed turret for the original four floors. (This cut costs and also gave them exposed decks that would later be used to mount equipment.) The basics, however, remained the same. The primary design goal was to make the building a Faraday cage—a completely enclosed conductive structure that would protect the occupants from lightning.

Workman's pride and joy was the rotating cupola atop the lab. The parts of the cupola, including the giant ring gear at its base, had to be built at the TERA machine shop. The idea was to enable cameras and other instruments to be rotated to track moving storms and other phenomena. A matching cupola was built for the tower of the Research Building on the Tech campus. The Langmuir cupola was rotated only in 1964 and 1965, when Brook mounted cameras and microwave antennas in and on it while studying the electromagnetic spectrum of lightning. Shortly thereafter it was put out of commission by dirt in the large bottom bearings. Repairing the cupola would have been expensive and difficult, and the rotation feature was no longer needed for any specific project, so the cupola was left in position. However, the big knife switch tempts idle hands to this day.

Charlie Moore gives the thumbs-up to a chopper pilot during the construction of the cable across Sawmill Canyon. New Mexico Tech Archives.

Construction of the lab began in the summer of 1962. Kenny Sorensen was the chief construction engineer for the project. This time, a local contractor was hired for the actual construction, but Workman was still saving money. Most of the materials were military and other government surplus, obtained free or for the cost of shipping. Workman would browse through TERA's storage yard for Army surplus, searching for Good Stuff, and say, "I'm not stealing this, I'm merely liberating it."

Workman had not spent much time on the site during the construction of the road, since roads were not really in his field of interest. But once the construction of the lab began, he made his presence felt. The statement he had made when the Research Building at Tech was being put up—"I want to approve the hammers that drive every nail into the wall"— turned out to be a good description of how he ran the Langmuir project. Sorensen was in charge, but everything he ordered and did had to carry Workman's approval.

A few problems did arise. One of the workers rolled his four-wheel-drive Scout off the mountain; he survived, but the Scout was mortally wounded. Someone found a fawn with a broken leg and took care of it, sending Sorensen into town for a baby bottle with which to feed it.

When winter arrived the crew quit work, as South Baldy is no place to be doing outdoor work in the winter. Since the bottom portion of the lab, the living quarters, had been completed, Workman looked for a live-in watchman. He found Floyd Reynolds, a survivor of the Bataan Death March. Reynolds also managed to survive winter on Baldy. His home in the lab was self-sufficient. Rainwater was collected on the roof and stored in a buried railroad tank car (surplus, of course). It was then pumped back up through a filter and gravity-fed into the lab. A diesel generator provided electricity. Reynolds brought his wife and a winter's provisions and kept an eye out for roving vandals.

But the hoodlums never got around to snowshoeing up the 10,000-foot peak. The lab was finished during the summer of 1963 and was dedicated on the Fourth of July.

6

The Sorcerer's Apprentice

By the time the lab's construction began, Marx Brook had been at Tech for six years and was beginning to work his way in earnest up the professional ladder. In a university, this is done largely by publishing articles in peer-reviewed professional journals. Between 1960 and 1962, Brook published a dozen papers on lightning in the *Journal of Geophysical Research,* working in cooperation with Workman and a Japanese professor named Nobu Kitagawa. They photographed lightning, recorded the changes in electric fields associated with it, and used radar to try to correlate it with other phenomena. In the dry season, they pored over their data and wrote papers, concentrating on the fine structure of the lightning flash and on what that might imply about the electrical structure of the cloud.

Brook and his collaborators also looked at lightning, often from the top of Socorro Peak, the site of one of their radars. One stroke that Brook and Vonnegut saw in a distant thunderstorm in 1958 was quite unusual. The normal lightning flash is composed of several events—most notably cloud-to-ground leaders and ground-to-cloud return strokes. The return strokes are the bright ones that onlookers perceive. As a rule, the successive stages are so short that they cannot be distinguished by an observer. This time, for some reason, things happened so slowly that the return strokes and the upwardly moving streamers could be counted with the naked eye. They wrote in the *Journal of Geophysical Research,* "It is refreshing, in this age of data-taking machines and digitalizing devices, to be able to report a simple visual observation which appears to have scientific worth."

Brook's emphasis on visual observation, on occasionally sticking one's head out of the radar room and looking at a thunderstorm, was understandable; he was an avid scientific photographer at a time when many of his colleagues were enchanted by their electrical instruments. He still takes pictures of lightning when he can, but he now has much more company. The various phenomena observed at Langmuir are now photographed by everything from time-lapse cameras that take a picture every five minutes to a Fastax camera, meant for filming explosions, that unreels 6,000 frames per second with a screech like a dentist's drill.

Even after the construction of the lab, Brook continued working on Socorro Peak and in Socorro, as did many of his colleagues. They now had a permanent mountaintop lab started, but it was not going to do them much good for a few years until it was properly equipped.

At Langmuir Laboratory, as in atmospheric research everywhere, logistics was a major problem. Despite the attentions of Kempton and his crew, the path to the lab was still rough and rocky. The people who worked at Langmuir became experts at packing fragile apparatus so securely that it probably could have been shipped parcel post. They learned those skills one pickup truckload at a time.

But the real problem was the "learning curve," not the curves in the road. Loading the apparatus onto a truck, driving it up the side of a mountain, and putting it in the lab were only three small links in a long chain of events. For the most part, one does not pick up a catalog and order atmospheric physics apparatus. Much of the equipment at Langmuir is either "home-brew" or converted military surplus. Making your own equipment and converting military surplus both start with applied physics: what are you looking for, how much of it do you expect to find, and how can you measure it? Catch-22 smirks at the researchers who do not have the equipment to get the data they need to design the equipment.

The winters pass quickly in experimental atmospheric physics. Designs must be revised, and next year's money must be sought. This year's equipment may be ready by the year after next. The equipment progresses through design engineering, skilled assembly work, testing, and usually redesign and more testing.

Time, money, and more and more people become involved. Scientists watch the thunderstorm season come and go from their workshops as the problems multiply. Electronic equipment fails illogically and turns out to have been wired up improperly. The telephone rings. One of the area newspapers wants to send a reporter down. A rancher thinks the research work is causing a flood; another suspects it is causing a drought. Do you suppose you could spare a few minutes? Eventually a piece of apparatus is completed and is ready to be hauled up the mountain.

So went the early sixties. Marvin Wilkening was the chairman of the Langmuir Laboratory Committee from 1963 through 1969. A surprisingly modest and self-effacing man for all his achievements, Wilkening says that he got the job only because Workman was too busy to do it himself. (The remark also shows a certain understanding of Workman's personality.) Yet Wilkening was named as the co-principal investigator in the proposal to build the lab, and as chairman he took time away from his own interest in atmospheric radioactivity to oversee projects in cloud physics and lightning.

Wilkening was a veteran researcher by that time, having come to the School of Mines in 1948, fresh from his doctoral work at the Illinois Institute of Technology in Chicago. During World War II he had worked on the Manhattan Project. He started with Enrico Fermi's atomic-pile group at the University of Chicago, moved on to Oak Ridge, Ten-

nessee, where uranium 235 was being separated from uranium 238, then went to Richland, Washington, where plu-
tonium 239 was being purified for a different type of atom
bomb. His Manhattan Project days culminated in 1945 with
an assignment to Los Alamos. When the first-ever bomb
went off on July 16, 1945, at Trinity Site, Wilkening was
there.

When the Manhattan Project was disbanded, Wilkening
resumed his interrupted studies at Illinois Tech, earning his
master's and doctoral degrees in only three years. But after
living in Los Alamos, even shacky wartime Los Alamos,
Wilkening and his wife found the crowds and weather of
Chicago unpleasant. He knew that he enjoyed teaching and
also wanted to go into research. A phone call to Dr. John
Harty, then head of the physics department at the New Mexico School of Mines, showed him the way out of Chicago
and into a research professorship. He wasted little time in
accepting.

At the School of Mines, Wilkening was taken under the
tutelage of Dr. William Crozier, who worked at the fringes of
atmospheric physics. Crozier was primarily a meteoriticist,
a student of meteorites, which was an important and exotic
field in those pre-Apollo days. Crozier also studied the formation of aerosols, or suspensions of tiny particles in the
atmosphere; this field was of more immediate interest to
the atmospheric physicists, because a cloud is an aerosol.
Wilkening went with him to Mount Withington to study the
radioactivity of aerosols.

Wilkening's work in this relatively unexplored area involved the measurement of radionuclides to trace air move-
ments. These radioactive variants of common atoms are
present in tiny amounts throughout the air. Their radioactivity makes them findable with the Geiger counter and its
more sophisticated descendants. Since these radionuclides

are blowing in the wind along with other atoms, they can be used to trace airflow between thunderstorms, over mountain ridges, and so forth.

That research on Mt. Withington brought Wilkening into contact with Moore and Vonnegut in the mid-fifties. It happens that radionuclides often exist in the atmosphere as positive ions. Moore and Vonnegut were interested in any kind of charged particles floating around in the atmosphere because the separation of positively charged particles from negatively charged ones is what allows lightning to occur. With Wilkening's help, they studied the behavior of the positive ions of radon and other radioactive gases under the influence of the storms' electric fields.

A long-lasting personal and professional relationship resulted from this cooperation, though the work was more directly profitable for Wilkening than for Vonnegut and Moore, because ionized radon is not a significant factor in cloud electrification. Thus, when the time came to build a permanent mountaintop lab, it was only natural for Wilkening to become involved. He helped scout out the territory when the scientists were trying to decide between Baldy and Withington; then he suggested the layout for much of the road on Baldy.

As running the newly constructed lab began to take up more and more of his time, Wilkening inevitably began using the lab as a base of operations for his own research. One of his major projects, done with the assistance of graduate student Dave Rust, involved tracing the airflow over and around the mountain ridges. A box of anemometers left over from that project still remains at the lab. It is practically a law of physics: a body that has been moved up to Langmuir tends to remain at Langmuir. Moore sometimes gets a mad gleam in his eye and hauls off a truckload of apparatus that has outlived its usefulness. But Wilkening's anemometers,

Langmuir Lab in 1964, brand new and uninstrumented. New
Mexico Tech Archives.

dusty, cobwebbed, some eviscerated for other projects, are still there. One never knows when something like that might come in handy.

Even when Wilkening's projects were not actually conducted at Langmuir, he benefited from the work done there, and the lab benefited from his research as well. That kind of interaction is one of the factors that makes the lab effective. A single flight over the mountaintop might take air samples, record wind speeds, and drop puffs of aluminum "chaff" to reflect someone else's radar. In the mid-seventies, visiting French scientists shot rockets to trigger lightning for American projects, and, in turn, shared the American data. A student driving a jeep up from Socorro makes the rounds of the Tech campus to see if anything needs to go up the mountain. In Langmuir parlance, this is known as "involving people in the decision loop."

As the sixties progressed, the decision loops at Langmuir grew larger and more complex. The biggest single change came in 1964 when Workman retired. (His retirement was short-lived; he soon went to the University of Hawaii at Hilo and founded a cloud physics observatory there.) The man who had rebuilt New Mexico Tech in his own image was replaced as president by former Lawrence Livermore National Laboratory scientist Stirling Colgate.

Colgate, first and foremost an astrophysicist, served as president until 1974, when he resigned in the midst of a personal scandal. He was not a strict administrator. Where Workman had maintained stern discipline, Colgate allowed virtually any kind of personal behavior.

For better or worse, Colgate's presidency changed the character of the school, especially insofar as the students were concerned. Workman's reign had been marked by strict yet unwritten rules of conduct. Once, some students were caught drinking beer in their parked car and throwing the empty bottles out the window. Workman impounded their car for the rest of the semester. But although Workman restricted personal freedom in some ways, he enhanced it in others. He provided freedom from having to lock the doors at night, for example, and freedom from having to worry about cheating on examinations. There was never any formal honor code; honor was practiced but not preached in Workman's day, when certain things were simply not done by gentlemen. By 1964 that attitude was already a dusty anachronism held in place by the sheer force of Workman's personality. To blame its demise solely on Colgate, as some have done, is to grossly oversimplify a complicated situation.

Although his reputation was made in astrophysics, Colgate was interested in many aspects of the physical sciences. Soon he gravitated toward the new atmospheric physics facility. Both directly and indirectly, Colgate was responsible for much of the progress that was made at Langmuir during the late sixties and early seventies.

But Colgate was not alone. Charles Moore, a de facto member of Tech since the mid-fifties, left the Arthur D. Little Company in 1965 for a position on the Tech faculty. Colgate had ideas for things to build; Moore has always loved to build things. Between them, the lab was in for a great period of growth.

The boom times at Langmuir started in 1966. By that time, the original lab building was getting rather crowded. Between the machine shop and the living quarters for the resident scientists, there was precious little space for research equipment. Moore and physics professor William Hume wrote a proposal to the National Science Foundation that year, asking for money to build an annex to the lab. Their work was a masterpiece of proposal writing. One of the illustrations was a glowing pen-and-ink drawing, an architect's conception of a second lab building. Another illustration, a black-and-white picture of dilapidated instrument trailers beneath a bleak gray sky, might as well have been a woodcut from Oliver Twist. The proposal netted $117,000 worth of sympathy from the NSF to erect the research building that came to be known as the Annex.

Other buildings went up on South Baldy at about the same time. The helium-filled balloons that were so important in the research work were being damaged or destroyed at their moorings when storms passed by in the night. Deflating the balloons each night and then re-inflating them in the morning would obviously have been impractical. A balloon hangar, erected in 1968, solved the problem. Also helpful was a truck-mounted balloon winch acquired from the Navy. The scientists had been carrying balloons around with a surplus DUKW amphibious vehicle, or "duck," that happened to have a winch on top. (Until its removal in 1984, the duck sat near the lab at an elevation of about 10,400 feet, which is probably a record for such a vehicle. Although retired from research for several years, the duck had remained as an invaluable conversation piece and the inspiration for innumerable flood jokes.)

But the most scientifically significant construction projects of the Colgate years had to do with cloud electrification. In 1960, in the cornfields of Illinois, Vonnegut and Moore had released space charge from elevated high-voltage wires. Airborne measurements had proven that the space charge was being carried up at 1,000 feet per minute or more when convective clouds were overhead. Unfortunately, none of the clouds they tried to charge up ever developed into thunderstorms. Similar experiments on Mount Withington had been unsuccessful because the wires, strung across the canyons, had been destroyed by the storms.

Now that Langmuir Lab had been built, Moore and Von-

Handling a balloon on South Baldy, circa 1965. New Mexico
Tech Archives.

negut had a grand opportunity to continue these experiments on a much larger scale, with sturdy apparatus, in a place where storms were sure to hit. They needed more than just successful partial results; they needed concrete data, backed up by extensive measurements, that could be obtained time and time again.

The first device built at Langmuir for this purpose was "Stirling's cloud machine." The machine consisted of a barnlike structure with a 1,200-horsepower airplane engine and propeller in one end and a chimney on top. When a storm passed over the ridge between Baldy and the lab, the engine was fired up and lubricating oil was injected into its exhaust. This produced clouds of smoke that were blown past a grid of high-voltage electrodes. From out of the chimney came billows of greasy space charge headed for the cloud at a high velocity. (The Forest Service was not informed of all the oily details of how the machine worked.)

At first, Stirling's cloud machine did not have a name, but an irreverent student took care of that. To ensure that the electrified smoke all went into the cloud, Colgate developed a sheet-plastic sleeve that would unroll like a party favor when hit by the air blast from the airplane engine. The student took one look and bestowed upon it the acronym of PBC. Visitors speculated about what PBC might stand for. Particulate Burner/Charger, perhaps? When they saw it in operation, they realized that PBC stood for Paul Bunyan's Condom. Colgate's device resembled nothing so much as a hundred-foot inflatable prophylactic.

Alas, the 1,200-horsepower contraption was not rigid enough to stand erect in the winds of a thunderstorm, so it was removed and the smoke was allowed to blow straight up from the chimney. Instruments for measuring the electrical effects of the smoke were installed on a cherry picker next to the machine. When Moore went up into the smoke

"Stirling's cloud machine" minus the PBC. New Mexico Tech Archives.

The PBC in action. New Mexico Tech Archives.

cloud on the cherry picker, sparks would fly from his eye-brows. Although both versions of the machine put out impressive amounts of space charge, there was no way to account for the role of ground effects in the results. The scientists could not tell for certain what effect, if any, their cloud of smoke had on the thunderstorms; Moore suspects that it had about as much effect as a blowtorch would have on an iceberg.

In 1968, with the aid of NSF money that came with the grant to build the Annex, the scientists went back to the original idea of releasing space charge from an elevated wire. This time, though, it would be done right. A little more than a mile of high-strength, half-inch-diameter steel cable was strung between a tower near the lab and another tower on Timber Ridge, two kilometers away on the other side of Sawmill Canyon. Since no road could be built along the top of Timber Ridge, the materials for the tower had to be taken to the site with pack mules and helicopters. After two summers of hard work in rough terrain, the cable finally hung across Sawmill Canyon. Made of special high-strength steel, it was suspended at each end from a six-foot insulator under 6,000 pounds of tension. It could be electrified with up to 150,000 volts when a promising cloud came near.

The cable produced some spectacular sparks. It also produced an interesting bit of lab folklore when the orange aircraft markers on it slipped down to the bottom of the catenary curve, and graduate student Larry Boucher went out on the cable in a bosun's chair to re-anchor them.

Unfortunately, the electrified cable produced such large local electrical anomalies that its effects on the clouds were obscured. Many investigators complained bitterly about this, so further cable experiments were postponed until an instrumented airplane capable of penetrating the clouds for direct measurements could be obtained.

The cable fell down during an ice storm in 1981. Over a mile of it lay in the bottom of Sawmill Canyon, wrapped around sixty-foot trees and needle cliffs. It was retrieved, with great difficulty, during the summer of 1984. By then an airplane known as SPTVAR I, the Special Purpose Test Vehicle for Atmospheric Research, was instrumented and ready to fly, and good results finally began to come in.

Other important additions to the lab were made in the late sixties. A spring in the bottom of Sawmill Canyon was equipped with a dam and a pump, and a water line to the lab, some 2,000 feet above, was installed. In 1969 came a three-centimeter radar on a tower atop the Annex. The radar, meant for cloud physics studies, scanned the entire sky vertically in horizon-to-horizon wedges like the segments of half an orange.

At about the time of the cable project, Colgate was beginning to drift back into astrophysics. He was and is interested in supernovae, the cataclysmic outpourings of energy at the collapse and death of very large stars. Supernovae are so rare that astrophysicists must search other galaxies for them. This is laborious and time-consuming work. Colgate had a brainstorm: why not build an automated telescope on Baldy that would check all of the visible galaxies several times each night?

That was easier said than done. The job involved automatically pointing a telescope very precisely at each of several thousand galaxies twice a night, measuring their brightness, comparing the brightness to a previously recorded value, and marking any great variations. It was a plausible idea, because a galaxy with a supernova in it is about twice as bright as usual for a few days. But building the automated telescope proved to be a phenomenally complex task.

Colgate started with the steel liner from a Titan missile silo. Atop the liner he built a clamshell housing for the tele-

Stirling Colgate (center, between horse and man with camera) gives a tour of the cloud-charger facility. New Mexico Tech Archives.

scope and its aiming apparatus. Once again, a giant phallic symbol reared up a few hundred feet from the lab. Colgate encountered the usual difficulties of erecting a large structure on top of a mountain, but the really hard part was building the electronic controls for the telescope. In fact, this proved impossible for many years.

Today, anyone building such a device would start with a microcomputer. In 1968 there were no microcomputers. Even integrated circuits were in their infancy, and Colgate did not have any. He set out to revolutionize astrophysics with discrete transistors and student labor. It took more than 15 years and many developments in electronic equipment before he finally succeeded. By that time, a team of researchers at the California Institute of Technology in Pasadena had built such a telescope—but they readily acknowledged that Colgate had provided the theoretical basis for their invention.

Of course, the conditions at Caltech in the early eighties were somewhat different from those at Langmuir in the early sixties. For example, Wilkening has a bearskin rug in his home. He shot the black bear himself after it ventured into the lab building one day in 1968.

With all the human activity on Baldy in the past twenty-five years, most of the bears have packed up and gone, as have the mountain lions. But the mountain is still the home of coyotes, a red fox, and George the Squirrel. George, who appeared in 1982, was the pet of Langmuir balloon chief Jack Cobb. "We need him to do our business," he would say as the squirrel ate junk food out of his hand. Moore would occasionally stand behind a tree and make clucking noises like two nuts being clicked together, trying to get George the Squirrel's attention. But the squirrel preferred the parade of people with food in their hands.

There were hazards besides the animals and the ever-present danger of lightning. John Rieche's Turn marks the spot where its namesake barrel-rolled a four-wheel-drive truck and lived to tell about it. Several years later, a riddle was coined about Langmuir "electroniker" Jerry Longmire. The riddle asked, "What's white and lies on its back with all four wheels in the air?" The answer: Jerry Longmire's jeep, which he had parked without chocking up the wheels.

In later years, these vehicles would be joined in the TERA boneyard by a new Chevy Blazer (flipped, somehow, on one of the flattest parts of the road by an astrophysics graduate student, who had a curve named "Adrian's Adios" in his honor) and by an old Dodge pickup that Colgate's wife forgot to chock up. No lives were lost on the Langmuir road until 1985, when a visitor died in an alcohol-related accident, but close calls have been all too frequent. The fact that no lab personnel have been killed or seriously injured on duty stands as testimony to excellent safety precautions—and to luck.

7

The Dream Thunderstorm

During the late sixties and early seventies, several new atmospheric researchers joined the New Mexico Tech staff and began to make important contributions to the Research. Among them were Bill Winn, an atmospheric physicist, and Paul Krehbiel, an electrical engineer.

Winn arrived in 1970, just after the facilities and activities at Langmuir began to grow in earnest. His primary work was to be the study of cloud electrification with airborne instruments; one of his first contributions was the use of instrumented rockets for probing thunderstorms. A variety of rockets have been used over the years at Langmuir, but Winn's favorites were military-surplus 2.75-inch attack rockets. In Vietnam, these rockets were carried by helicopter gunships for use against enemy troops. Winn put them to a more peaceful use, replacing their explosive warheads with electric-field meters and radio telemetry circuits. When the storm overhead grew interesting, Winn would push the button. The rocket would shriek off at supersonic speeds, penetrating the cloud and sending back data.

Although their warheads had been removed, these rockets were still dangerous. They were two feet long and fairly heavy, and they went off like—well, like rockets. Restricted Airspace R5113, the only restricted airspace ever set up for civilian research, was established over the lab area in 1971 at the request of Winn and Moore. That took care of the danger of shooting down stray private pilots.

Krehbiel, as an electronics engineer, was not involved in hypothesizing so much as in designing instruments to gather data. One of his most important contributions to the re-

The Annex under construction and (*right*) just after completion.
New Mexico Tech Archives.

Marx Brook on a visit to the lab in 1986. Joe Chew photograph.

search effort was the Fast Scanning Cloud Physics Radar, which Marx Brook and he invented. The unit, patented in 1975, was soon nicknamed the Redball Radar because of the spinning red ball that housed its antenna. It transmits broadband radio-frequency "noise" in order to explore various aspects of a cloud's structure rapidly. Mounted on a semitrailer, the Redball Radar has been towed up to South Baldy for several summers. It is getting old, at least by electronic standards, and its ball is now white rather than red, but it continues to provide useful results.

In the mid-seventies came a pair of aircraft, both powered sailplanes, for the lab's use. One of them, equipped with an exceptionally powerful engine and designed for high-altitude flight, was assembled specifically for use at Langmuir out of parts from various sources. The plane has "NOAA" (for National Oceanic and Atmospheric Administration) painted on the fuselage and "Air Force" painted on the wing. The fuselage used to read "Navy." The plane's mixed parentage is evident only to people who go to the Socorro airport and look inside the Tech hangar, for it has not flown in several years. It needs a new engine, which will be purchased and installed as soon as someone comes up with an extra $70,000 or so.

The other plane, SPTVAR I, has been more useful. The Special Purpose Test Vehicle for Atmospheric Research is a powered Schweizer sailplane that began its career as a reconnaissance drone in Vietnam. After the war it went to the Office of Naval Research (ONR), where the remote-control equipment was removed and a rather cramped space for a pilot was carved out. Reinforcements were added to permit relatively safe flights through active thunderstorms. In 1975 the ONR provided the plane to Tech on a long-term loan. It became the domain of veteran research pilot Bill Bullock, formerly chief pilot for the National Center for Atmospheric

Research. Bullock now runs Air, Inc., in Colorado Springs, Colorado, and flies SPTVAR I during the summers.

An old saying has it that there are old pilots and bold pilots, but no old, bold pilots. That saying is useful to keep in mind as Bullock, who has made it to his early sixties unscathed, buzzes the Annex on a flight down Sawmill Canyon. (An apocryphal bit of lab folklore holds that when the balloon hangar was first built, Bullock used to fly over it low and slow and ever-so-gently brush the roof with his landing gear.) The former Air Force Academy flight instructor has never had an accident in more than 10,000 hours in the air.

His association with Moore goes back many years; he collected the electric-field data from Moore's 1960 experiments with charge release from cables suspended over the Illinois flatlands. With Wilkening on board, Bullock has also flown into dicey situations in aircraft not built for thunderstorm penetration and handled the danger with complete aplomb. Considering his experience and safety record, Bullock, it would seem, is no more of risk-taker than any other research pilot; he merely enjoys demonstrating his skill before an appreciative audience.

All these planes, radars, and other instruments help scientists study what is going on inside a storm, where the winds and electric fields are high, the visibility nil, and the physical processes still poorly understood. Not everyone comes into contact with these conditions as intimately as Bullock, whose plane is regularly bombarded by hail, scorched by lightning, and tossed around like a feather by 5,000-foot-per-minute updrafts and downdrafts. But thunderstorm research calls for similar levels of skill in many less spectacular disciplines. The work rewards genius and demands at least a very high level of competence.

Thunderstorm research also seems to attract colorful individuals, such as Workman, the iron-willed tyrant for whom

Growing cumulus and a mature anvil cloud. Joe Chew
photographs.

Day lightning in Sawmill Canyon. Joe Chew photograph.

Night lightning. Bill Winn photographs.

Rain shaft near the laboratory. Bill Winn photograph.

Rain and lightning from a summer thunderstorm. Bill Winn
photograph.

many people would have walked through fire; Brook, who is like Workman with a louder voice and a noisier sense of humor; and Colgate, radically different in character from Workman but perhaps even more ingenious. Today, as throughout the seventies, the character of Langmuir Laboratory and the work done there is shaped by Charlie Moore.

Given his live-wire personality and his self-aware importance at the lab, it seems natural that the reporters who visit from time to time should gravitate toward Moore. He claims to dislike journalists, those annoying people who waste his time and mangle the facts. Nonetheless, he is the featured subject of an astonishing number of photos and news films shot at the lab. When a journalist begins to fiddle with a camera, a change comes over Moore. His eyes light up and his crewcut stands at attention as he waves his arm over the vast expanses of terrain or hush-puppies his way up the guy wires on the cable tower. When he gets caught up in the excitement and attention, he can even manage a blush.

One of the first things people notice about Moore, along with his balloonist's jumpsuit and military brush-cut, is the scientific flavor of his speech. To express similarity, he is apt to say "on the order of" rather than "more or less." Sometimes he "repairs" things, other times he "reduces their entropy." But it goes beyond the use of such conversational artifacts. Moore's language is colored by the jargon of physics because he thinks in terms of physics.

There is no telling when or where physics might enter a conversation with Moore, but it almost always does. For example, on the morning ride up the mountain, with Moore at the wheel of the crew-cab pickup, a casual mention of golf may turn into an explanation of why golf balls have dimples. The reason, of course, is that dimpled ones fly farther. This leads into a discussion of laminar versus turbulent flow of a fluid over a surface, and of the Reynolds number. The dimples create micro-vortices at the surface of the flying ball. This increases the Reynolds number, which means that the drag of the ball is lowered. Common sense dictates that the skin of a high-performance aircraft should be smooth. Aircraft designers, who are just now getting the computer capability to replace common sense with mathematics, have begun contemplating dimpled surfaces. Moore has gone from the clubhouse tee to the aerospace industry in just a few miles, using the rearview mirror to make eye contact with his backseat passengers.

Under the dashboard of the truck is a two-way radio tuned to the Langmuir frequency. Moore bought it because the satchel-pack walkie-talkie he used to carry did not have enough power to reach the lab from the highway. He takes the truck home at night, but not to listen to the radio; he has another Langmuir radio in his bedroom.

One of Moore's perpetual companions at the lab is thin, chain-smoking Dr. Charles Holmes, known as Doc Holmes. The nickname helps reduce the confusion caused by having two Charlies at the lab, but it also fits him. While Moore could pass for a retired Air Force jet jockey or an aging construction engineer, Holmes, who came to Tech in 1959, is clearly a scientist. An instrumentation physicist, he has the bloodshot eyes and stooped posture that come only from years at an electronics bench. He is as reticent as Moore is garrulous.

Doc Holmes is one of the best and fastest drivers at the lab. Hunched over the wheel with a crooked grin on his face, he sails down the highway with blissful disregard for the speed limit. On the mountain, he slows down a bit, but seems to have complete confidence that the small lateral excursions of the vehicle's rearend have no effect on its course. It is entirely possible that he saves wear and tear on his tires by hitting only the high spots on the road. Another hypothe-

Checking out some apparatus in the newly completed Langmuir
Lab. New Mexico Tech Archives.

sis, advanced by Langmuir balloon chief Jack Cobb, is that Doc Holmes's tubeless tires go soft all the time because he literally beats the air out of them.

Holmes, of all people, was the president of New Mexico Tech during most of 1982 and part of 1983. It was an interim presidency, a mission to maintain the status quo while the Presidential Search Committee sought a permanent replacement. His predecessor, theoretical physicist Kenneth Ford, had resigned in the midst of managerial scandal at the urging of the school's Institute Senate. "Dump Ford" was a popular campus slogan often parroted by people who did not have the slightest idea why Ford should or should not be dumped. It seemed that all the frustrations of the students, faculty, and staff had found a single outlet: Ken Ford.

The last straw for Ken Ford was the construction of Macey Center, a $6.5 million theater and conference center overlooking the Tech golf course. In spite of its award-winning architecture, the building was immediately dubbed "Ford's Folly." Even the building's critics had to admit that it was beautiful, especially when compared to the Quonset hut that had been used as the school theater. But it created a formidable financial problem. By the time the center was dedicated in April 1982, the school had exhausted its cash reserves and was deeply in debt. A freeze on wage increases and hiring was imposed while Tech awaited an emergency bail-out from the state legislature. Only the R&DD, with its large government grants and contracts, remained in the black. These were the circumstances under which Doc Holmes assumed the interim presidency.

Holmes did not seek the position; he merely failed to run fast enough when it was thrust upon him. On the surface, he seemed to be an ideal interim president, a figurehead who would passively allow his advisors to run things: shy, unassertive, more at home in front of an oscilloscope than behind a podium. Wilkening once observed that anyone who is that good at electronics has to be an introvert. President Holmes once left a particularly hectic regents' meeting and went straight to his laboratory in the tower of Workman Center, locking the stairway door and turning off the power to the elevator. From his stooped walk to his absent-mindedly professorial demeanor, he looked like the kind of president that power-hungry subordinates dream of.

But appearances were deceiving. Old-timers at Tech have fond memories of his term as president of the Institute Senate, where in half an hour he would dispose of business his predecessors would have spent all afternoon on. Newcomers would occasionally try to bully the "meek" little physicist, only to find themselves staring into a pair of protuberant, bloodshot blue eyes with all the warm vulnerability of glass. When Doc Holmes gave them "the look," strong and ambitious men suddenly developed the urge to sit down and shut up.

That side of Doc Holmes occasionally comes through at Langmuir. The instruments at the lab proper—not the kivas, but all the radars and field mills and the far-flung networks of rain gauges and thunder mikes—are his babies. He designed most of the electronic systems at the lab and built many of them himself; now he maintains them as well. He comes up in the morning, puts down his sack lunch, gets a cup of coffee, and goes to work. It is solitary labor, even more so than most electronics work, for Holmes would rather do something himself than explain to someone else what he wants.

Occasionally, though, Holmes will delegate responsibility to a student assistant who is handy with electronics. Sooner or later, that student will get "the look." It is inevitable, because the schematics for a dismaying amount of Langmuir equipment are kept in Doc Holmes's head and nowhere else.

Holmes will respond amiably enough to intelligent technical questions. A stupid question, though, or one that breaks his concentration during an all-morning trek through the innards of an instrument, will cause him to glance up through a cloud of cigarette smoke, soldering iron still in hand, and say, "Read the damn manual." Then he turns back to his circuit.

The staff and visiting scientists at Langmuir are trained and intelligent people who are there to do a job. Consequently, there are very few rules: don't drive off in other people's jeeps without permission; don't leave garbage where bears can get it; try not to start forest fires. Violations are forgivable, although a particularly colorful incident might go down in lab folklore with the culprit's name permanently attached. The lab could never have been built if the people in charge had not had a sense of humor.

But it is something different to damage an instrument through carelessness, neglect, or simple damn-foolishness. That brings a heated chewing-out from Moore, plus a draft of icy, disappointed scorn from Doc Holmes. The recipient of all this attention is not likely to make the same mistake again.

By the mid-seventies, these people and facilities had made the lab an entity to be reckoned with in the world of atmospheric research. The instruments had been built, the personnel had been hired, the techniques had been polished. Langmuir Laboratory, ten years old and on the upslope of the learning curve, was ready to go to work. The only thing that remained was for nature to cooperate. Some day soon there would come the wettest, stormiest summer in living memory, and all the instruments everyone wanted to use would be in place and working at the same time. It would all come together like a dream . . .

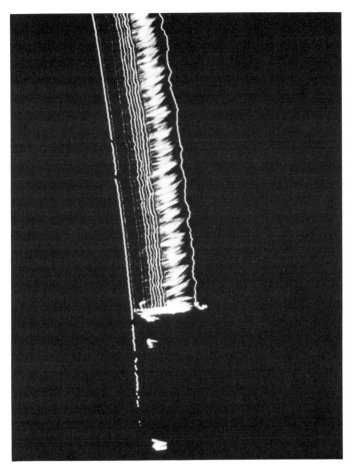

Triggered lightning at Langmuir Lab. The multiple streaks in this time exposure are various components of the lightning flash following the conductive channel as it is blown by the wind. Ray Nelson photograph.

The temperature is already in the eighties on a sunny morning in Socorro as Moore, Holmes, and the other daily commuters leave the Tech campus in their pickup truck. By 9 A.M. they are almost up to the top of the mountain, which lies beneath the shadow of the first faint wisp of cloud. The commuters pull up at the lab, discuss the order of the day's business with the people who stayed overnight, and scatter to the various sites to prepare for the coming storm.

By noon the early-morning gossamer cloud has turned into an ominous cumulonimbus. Sandwiches and coffee are wolfed down in front of chart recorders and radar scopes as the storm develops—and everything works! For an hour, an hour and a half, *two* hours the data pour in. The radars are all up and running, the field mills are monitoring the electric effects of the cloud, the planes are in the air. Doc Holmes's thunder mikes, the sixty-hertz AC hum exorcised from their signal lines only hours before, are pulling in the separate cracks and booms from every zigzag in the lightning strokes. Brook's and Moore's staff gather data that will give them something to argue about for the next ten years. As the cloud grows old and the storm calms down, scientists take a well-earned coffee break and think about the coming winter, when their graduate students—the best bunch anybody can remember—are going to analyze all the data.

. . . somebody honks a horn and the dreaming scientist wakes up. Three o'clock in the morning. Better get back to sleep. It's going to be a long day of malfunctioning apparatus, hard physical work, and maybe, just maybe, a forty-five-minute shot at a puny thunderstorm that might blow over to Baldy from Withington.

One of the decade's biggest events at Langmuir was TRIP, the Thunderstorm Research International Project. TRIP research was conducted at the lab in 1979 and 1981. Among the most important of the TRIP participants, at least from the point of view of the Langmuir scientists, were the French. They brought along a well-developed technology for causing lightning to strike when and where the scientist wanted it to.

Actually, the basic technique of triggering lightning was more or less an American invention. The seminal paper on the subject, "Artificial Initiation of Lightning Discharges," was published in November 1961 by Brook, Moore, Vonnegut, and a pair of co-authors who helped them get access to a van de Graaff generator.

It all started when the Navy tested a depth charge in Chesapeake Bay one day in the late fifties when there was a thunderstorm overhead. Since water is virtually incompressible, depth charges set off at moderately shallow depths throw a great gout of water into the air. In this case, the spray reached a level of seventy meters above the sea in one second and caused lightning to strike.

In the meantime, Vonnegut and Moore had been using balloons to carry long wires into thunderclouds in early attempts to electrify the clouds. When the clouds turned into thunderstorms, lightning would often hit and burst the balloons, but the wires themselves were never hit. Since large electrical currents flowed in the elevated wires, Moore and Brook speculated that the emitted electric charge formed a protective sheath around the wires. The next question was, would lightning strike a wire that was moving through the air so fast that the sheath of electric charge could not keep up with it?

To answer that question, Vonnegut, Moore, and their colleagues set about using the van de Graaff generator at the Boston Museum of Science to try to repeat the phenomenon under controlled conditions. They stretched a wire beneath the van de Graaff, which is a high-voltage static-electricity generator, and fired up the machine.

First they conducted the experiment with the ends of the

wire fixed in place. The length of wire began to oscillate like a jump rope. The wire's oscillation and the lack of lightning indicated that corona current, a low-level continuous discharge, was passing from the van de Graaff generator to the wire and thence to ground.

In the next phase, in order to prevent corona current from ruining the experiment, the scientists used a giant elastic band to make a kind of slingshot; it would snap the wire into the vicinity of the van de Graaff faster than the electric field around the wire could build up. This arrangement produced a lightning-like discharge every time.

Their report on the project suggested ways of carrying a wire into the electric field below a thunderstorm at the necessary speed. The scientists suggested bows and arrows, as well as balloons and kites equipped with elastic. Last but not least, they suggested the use of rockets trailing wires.

It was not a new idea. The first attempts to use wire-trailing rockets to trigger lightning had been made in 1753 by a scientist named Giambatista Beccaria. But the technique remained a footnote to scientific history for more than two centuries until Moore, Vonnegut, Brook, and the rest revived it. With their paper in mind, Morris Newman, working in the early 1960s, aboard a ship off the Florida coast, had great success using wire-trailing rockets to trigger lightning.

Despite Newman's work, the technique languished for several more years. Numerous experimenters attempted to use rockets to trigger lightning on land; nearly all of them failed. (The Apollo 12 rocket triggered lightning shortly after liftoff. Although the incident yielded no physics data, it sent shivers down the spines of onlookers, some of whom knew firsthand what it is like when a booster rocket explodes on the pad.) The problem with most of the triggering experiments

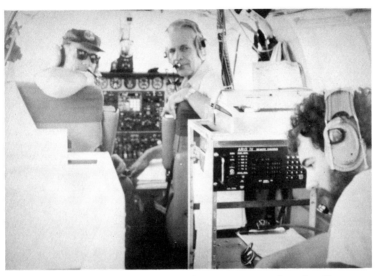

Marvin Wilkening (right) and flight crew prepare for takeoff.
New Mexico Tech Archives.

was the equipment; most of the experimenters used small military rockets that accelerated so fast the wires broke.

While the Americans were triggering lightning with Saturn 5 boosters (and failing to do so with antitank rockets), the French scientist Dr. Pierre Hubert was busy working on another elusive technology, controlled thermonuclear fusion. But in the early 1970s Hubert turned his attention to lightning and thunderstorms. In 1972, he and his associates proposed a technique for triggering lightning with rockets without breaking wires, and in 1973 they tested it successfully at an atmospheric physics laboratory in France.

The basis of the French technique was something Workman and Langmuir would have appreciated: a paragrele or anti-hail rocket about one and a half feet long. Atop the blue plastic rocket was a yellow nose cone intended to contain silver iodide and a small explosive charge. Setting off this charge inside a thundercloud would supposedly suppress hail. Actually, according to modern theories, the explosion would encourage hail formation if it did anything at all, but Hubert was not interested in that. He removed the silver iodide and explosives and added a wire connected to a dispensing reel intended for wire-guided antitank weapons. The reel stayed on the ground while the rocket pulled the wire into thunderstorms. With luck, lightning would travel along the wire. (The wire does not act as a lightning rod, intercepting a lightning stroke in progress; rather, it causes a stroke that otherwise would not have happened.)

Hubert and his colleagues perfected their technique in France and brought it to Langmuir in 1981 and 1982. There the thunderstorms were much more frequent than at the best sites in Europe. Also at Langmuir were other experiments that could benefit from reliable lightning triggering. Langmuir scientists have since developed their own triggering system, which uses smaller hobby-type rockets but is otherwise similar to the French method. After 220 years, artificial initiation of lightning had finally become a proven technique.

As the years and the storms went by at Langmuir, rain gushes occasionally occurred shortly after lightning strokes. In their attempts in the late fifties to study the relationship of rain to lightning, Vonnegut and Moore had been frustrated by the inadequacy of the available radars. They remained frustrated for twenty years, until a graduate student, with considerable assistance from Doc Holmes, built a suitable radar at the lab.

The first version of this S-band radar at Langmuir was put together by Anthony Atchley in 1977 as a master's thesis project. This unique radar, an agglomeration of military-surplus assemblies and homemade circuits, was designed specifically to test the lightning-rain relationship. It sends out a signal with a wavelength of eleven centimeters, which is not reflected to any great extent by clouds, but is reflected by heavy rain and by the plasma channel left behind by lightning. The unit's range was selected for coverage of the lightning-triggering area between the lab and South Baldy. The idea was to point the radar dish at the triggering site during each rocket firing until a rain gush was finally triggered. The radar would then show exactly what happened where and at what time inside the cloud. When Atchley finished his thesis project, it appeared that the question would soon be settled.

The enthusiasm looked premature, and it was. Working all the bugs out of the radar took seven years and at least one major design change by Doc Holmes. The problems were numerous. Atchley's original design was not quite ideal for the job. Various components of the radar deteriorated with age. The little brown moths that hibernate by the millions at the lab got into the waveguide, where they acted as signal attenuators. The radar finally began to yield some useful data in 1984. Moore reported one rain gush that was observed by radar immediately after a lightning stroke, but interpreting the data could take years.

Marx Brook was not standing idly by while the S-band radar was being built. He and some of his colleagues, including Krehbiel, spent three summers at the Kennedy Space Center in Florida in the late 1970s. There they observed thunderstorms and chased spiders out of their field mills. Coastal Florida has about the worse possible climate for scientific instruments. Humidity and bugs were the chief problems. Spiders, in particular, were to the Space Center as the little brown moths were to Langmuir. Every day the researchers had to remove spiders from their apparatus. By the third and final summer, they had finally figured out a way to build a spider-resistant field mill.

The scientists encountered another natural obstacle, thanks to the state Fish and Game Department. One day the Fish and Game people decided that a 600-pound alligator living in the area needed to be relocated to the Everglades. The physicists watched as the alligator was lured from the swamp with a sandwich, and a Fish and Game man slipped a noose over its head. The 'gator immediately headed for deep water and the noose slipped off. A second sandwich tempted the creature into sticking its head back on shore. The noose was slipped back on, and the alligator lost a tug of war with the physicists. Someone taped its mouth shut, a procedure missed by the Langmuir researcher who was filming the event. Krehbiel prudently provided extra reinforced tape to make sure the mouth would stay closed.

The next step was lifting the 'gator into the bed of a

Bernard Vonnegut (left) scores a point against Marx Brook at NASA.

Graduate student Barbara Zinn takes radiosonde data in the Balloon Hangar, 1983. Joe Chew photograph.

pickup truck. Four physicists lined up on each side of the beast. With considerable difficulty, they picked up the 600-pound animal. The torpid 'gator sprang to life and began thrashing its tail. They put it back down in a hurry. On the second try they succeeded in getting it into the truck. The Fish and Game men slammed the tailgate and waved good-bye. They never said who untaped the 'gator's mouth.

Somewhat more scientifically profitable for Brook were his trips to Japan in the 1970s. There he studied the famous winter thunderstorms of the country's northeast coast. Besides their ferocity and the fact that they occur in the dead of winter, these storms are famous for their lightning. Brook has published measurements showing that the strokes come out of positive charge in the top of the cloud, rather than out of the negative charge in the bottom as happens in most storms. The Japanese storms, according to Brook, are initially electrified through "conventional" processes—that is, not through convection. Moore, of course, will argue about that.

While Brook was chasing alligators in Florida and winter thunderstorms in Japan, Moore was digging a hole on top of South Baldy. Into the hole he put the first Kiva, a steel cylinder fifteen feet in diameter and ten feet tall, designed to protect equipment and people from lightning. As it was being lowered into the hole, the Kiva did not look very impressive. It resembled an underground gasoline storage tank, not an outpost on the frontiers of thunderstorm research. The Kiva was soon buried so its roof was flush with the ground. Then the inner parts of the Kiva were lowered through the roof hatch and installed.

The important parts were all electronic, for the Kiva is an instrumentation bunker. The Kiva took its name from the round, subterranean ceremonial rooms built by some of the

Jack Cobb releases a radiosonde on South Baldy, 1983. Joe Chew photograph.

Charlie Moore and student Dave Lowe install the "stinger" atop
Kiva II, 1983. Joe Chew photograph.

Triggered lightning at Langmuir Lab. The numbers were displayed inside the camera to indicate the time the photo was taken. Charles Moore photograph.

Native American tribes of the Southwest. The Langmuir Kiva is like a shrine to newer gods, high-tech deities with names like Nanofast and Lecroy and Hewlett-Packard. Inside, tucked away from storms and lightning, are digitizers, data recorders, and a number of small computers. Armored cables carry electrical impulses down from electromagnetic sensors on the surface.

The installation of the Kiva marked a subtle change in the focus of the research. The visible properties of lightning and its electrical effects on the environment had been studied. Now it was time to get underneath some lightning strokes and study their electromagnetic properties up close. The Kiva was installed in 1977. Within six years it would be joined by another, larger bunker, Kiva II, funded indirectly by the Air Force. As the seventies ended, the study of the electromagnetic effects of lightning was becoming an important area of atmospheric research.

8

Lightning Hazards

In the course of fifty calendar years and countless man-years of collective experience, the Langmuir scientists and their colleagues have become well-acquainted with the hazards posed by lightning. When a good stroke hits the metal roof of the lab cupola, it sounds like a bomb going off. When it hits a tree, the tree is usually charred; sometimes, if the current vaporizes enough sap and water, it explodes. Occasionally lightning strikes someone standing with upraised putter beneath a thunderstorm, and there is one less golfer in the world. Anecdotal experience with lightning goes back to the beginning of recorded history. But science demands more than anecdotes. Accordingly, recent years have seen a great increase in the systematic, quantitative analysis of what happens when lightning strikes an object or creature.

Myths abound concerning lightning strikes at ground level—that television sets attract lightning, for example—but it is in the skies that people are most afraid of lightning. Flying is, at best, a great expression of faith in technology; a good thunderstorm, with winds dangerous in their own right in addition to lightning, tends to erode that faith in air travelers who normally suppress their doubts. Looking out the window and seeing lightning in midair makes many people wonder what would happen if lightning were to hit the airplane.

And lightning does hit airplanes fairly often. Usually the strike goes unnoticed. Only the ground crews doing a routine inspection after the flight notice the damage, which is usually confined to small burn holes. (When a jet streaks through the path of a lightning flash, the successive components of the flash burn a neat line of pinholes in the skin of the plane.) Occasionally the damage is more spectacular. In the early sixties, for example, a 707 had part of its radome cracked by lightning. Other types of problems are possible. A pilot on final approach, for example, could be momentarily blinded by a lightning flash just ahead.

Although most plane crashes in thunderstorms are caused by wind, lightning occasionally plays a direct role. For example, lightning can burn through the metal skin of an aircraft; if it penetrates a nearly empty fuel tank that is full of fuel vapor and has not been charged with an inert gas, an explosion can result. Fortunately, such dramatic events are uncommon in metal-skinned planes. But two trends in the design of high-performance aircraft are increasing their lightning vulnerability.

The first is the use of plastic-like composites for the airframes. These materials are stronger and lighter than most common engineering metals. The catch is that they usually do not conduct electricity. Aircraft designers used to wrap their planes in metal and take it for granted that lightning would not get inside. Composites have forced them to think analytically about lightning and its behavior, because protection against lightning no longer automatically comes with the hull of the aircraft.

Second, the innards of the modern aircraft are more vulnerable to lightning. Avionics, or aviation electronics, has come a long way in a short time. The pilot of a state-of-the-art airliner sits before a computerized instrument panel that shows all the information a pilot needs on cathode-ray monitors. Sophisticated radar probes the skies ahead. A computer and a stack of radios allow precise navigation to every part of the globe. All of this equipment has to be hardened against lightning damage to a certain extent. But how can it be hardened? And to what extent? These are the questions that designers have to ask themselves, and man's knowledge of lightning is such that they often have to go to basic research for the answers.

Even the controls of the latest aircraft are electronic. An older airplane—a 707, say, or an F-4 Phantom—has control surfaces linked to the pilot's hands and feet by cables and hydraulic power boosters. Such linkages are virtually lightning-proof, but they are heavy and slow to respond. Enter the fly-by-wire aircraft, such as an F/A-18, maneuvered by servomotors, with a computer and a bundle of wires between the pilot and the control surfaces. Light. Lightning-fast. But also lightning-vulnerable.

The Grumman X-29, which became famous for its forward-swept wings, is the ultimate embodiment of these new trends. It is a totally fly-by-wire design, and is made mostly of composites. The X-29 will never be flown anywhere near a thunderstorm. It will probably never even be taxied on a wet runway. But with the advent of new technologies and materials, the government and aerospace companies are eager to find out exactly what lightning does to an airplane.

The theoretical side of that research brought Air Force Flight Dynamics Laboratory scientists to Langmuir three years in a row. Physicist Richard Richmond was one of the leaders in their effort to quantify the electromagnetic effects of a lightning strike on a fuselage. He designed LSO-1 and its successor, LSO-2, the instrument-stuffed Lightning Strike Objects. He and a few colleagues hauled these forty-foot aluminum cylinders from Wright-Patterson Air Force Base near Dayton, Ohio, to Langmuir Lab.

Once at Langmuir, the scientists suspended the cylinder inside a tripod of telephone poles. A battery inside the LSO provided power without the risk that lightning-induced surges would come down the power lines. To achieve even more isolation, the data from the instruments were optically encoded inside the cylinder and sent to a nearby trailer via fiber optics. Since fiber optics do not conduct electricity, there was no way for lightning to induce a surge on them. Richmond and his colleagues waited inside the trailer.

And waited. And waited. Their final year at the lab, 1982, was a disappointing thunderstorm year. On several occasions, however, the visiting French scientists managed to trigger lightning that struck the cylinder. Finally, late in the summer, a big thunderstorm rumbled up over the test site and lingered for an hour or more. Richmond, who was in Socorro that day, drove up during the storm and parked in a safe area. He watched in ecstasy as lightning struck his cylinder three times—twice after rocket fire and once all by itself. When the storm blew over, he rushed to the trailer, only to find that a power outage had disabled his data recorder.

The efforts were not a total loss, however. Richmond had good data from earlier storms at Langmuir, and also from storms at the Kennedy Space Center. He obtained the kind of data that would interest a physicist rather than an engineer. His goal was to develop a mathematical model of the electric and magnetic fields around a conductive cylinder, a shape chosen primarily for its mathematical simplicity. (Physicists choose spherical or cylindrical symmetry anywhere they can, because it makes their job so much easier.) The fact that the cylinder resembled an aircraft fuselage was secondary.

Other visitors to Langmuir have more pragmatic goals. The best way to test a supposedly lightning-hardened design is to trigger lightning onto it and see what happens. Although Sandia National Laboratories in Albuquerque, a major defense research lab, has learned how to trigger lightning, Langmuir is far more experienced in the field and has better facilities and a lot more thunderstorms. So a team of

Sandia engineers came to Langmuir late in the summer of 1982 with a "package."

The package came up the mountain early in the morning. Along with it came a dozen or so security officers. Most of them were in uniform and carried assault rifles. Their boss, who wore a suit, left his smile at home, if indeed he owned one. He was big enough to make a good defensive end, so he did not look awkward packing his .44 Magnum in a shoulder holster. The group took possession of the peak of South Baldy and set up their package and a machine gun on the summit. All morning, the students at the lab speculated about the contents of the package. Guidance electronics for a cruise missile? Avionics for an F-14? The arming mechanism for a nuclear weapon?

Few people knew, and they weren't talking. Security is tight at Sandia Labs, located next to Kirtland Air Force Base. The scientists who came with the package wore their security clearance badges at Langmuir. They were not happy about the presence of the French scientists at the site. (The two visiting scientists from the People's Republic of China who were living at the lab had been sent down to Socorro for the occasion.) Lightning was triggered upon the package that afternoon—a fine stroke, measured at seventy-seven kiloamps of peak current—and the Sandians went home. Nobody ever said whether the package had passed its test.

As a rule, the military presence at Langmuir is not nearly that blatant, but lightning research has many military applications. A constant reminder is Kiva II, built next to Kiva I at the behest of the Air Force and the Summa Corporation, an Albuquerque firm. Dr. Carl Baum of Summa Corporation personally inspected the lightning-proofing of Kiva II, right down to specifying what kinds of transient arrestors should be built into the power supply. Baum is a brilliant electrical engineer according to most opinions, including his own. He and Moore enjoy their arguments. The general understanding is that Moore is the final authority atop South Baldy, but that Baum has ways of clamping the Air Force money hose.

Kiva Two is essentially a bigger version of Kiva I. It is about twenty feet in diameter, round, with a steel roof flush with the surface of the peak. It has two walls of steel with a crawl space in between. The aluminum door is elaborately grounded. Above it, the top of South Baldy is covered with steel wire fencing, with grounding stakes every few feet. The fencing provides an electromagnetically smooth and highly conductive ground plane for the measurements made at the kivas; this becomes important at data-analysis time. The need for the elaborate grounding and transient-suppression schemes inside the kivas is more immediate. The kivas, which are manned throughout the storm experiments, are at the top of South Baldy. Kiva II has rocket launchers on its roof, and lightning is triggered onto a target next to the launchers.

Some of the work done at Langmuir has military applications aside from the obvious usefulness of triggered lightning in equipment testing. One of the effects of a nuclear explosion is a strong electromagnetic pulse, or EMP. Some scientists have speculated that a large hydrogen bomb exploded many miles above the center of the United States could disable non-hardened communications and the nation's power grid through EMP damage. Lightning produces an EMP—a much weaker one than that of an H-bomb, but good enough for research purposes at short distances. By studying the electromagnetic effects of lightning, defense researchers can get a better idea of what EMP does and how to harden military equipment against it.

Lightning hazards can arise in the most unlikely areas. After the Comprehensive Test Ban Treaty was signed in 1963, outlawing nuclear weapons tests in the atmosphere,

LSO-2. John Shortess photograph.

Triggered lightning strikes LSO-2. John Shortess photograph.

undersea, and in space, the United States needed a way to monitor Soviet compliance. An electro-optical system was developed to watch for the bright flash of an illegal nuclear explosion in space. Unfortunately, the system did not work properly, since it could not reliably distinguish a nuclear test from lightning. In 1964, Marx Brook was awarded a contract by the Defense Advanced Research Projects Agency to find a way to make the system more discriminating.

Brook started by studying the differences between a lightning flash and a bomb flash. Using the cameras and microwave instruments in the cupola at Langmuir, he investigated the spectrum of lighting and found that electrical discharges in the presence of water cause the hydrogen atoms in the water to emit light of a specific wavelength. It was also known that another unique wavelength is emitted by nuclear blasts, which provided a double-check. The lightning-filtering system Brook proposed after this research was tested in short order and was quite successful.

Hardening against the thermal and electromagnetic effects of lightning is also important to the explosives industry. Moore, along with Brook and physicist E. Philip Krider of the University of Arizona in Tucson, published *A Study of Lightning Protection Systems* in 1983. The report resulted from a multi-year study funded by the Defense Munitions Board. The object: to make recommendations for better lightning protection for ordnance plants.

The possibility that an explosives plant might be blown off the map by a lightning stroke has long been recognized. In 1876, in a report to the British Association for the Advancement of Science, physicist James Clerk Maxwell wrote, "What we really wish is to prevent the possibility of an electric discharge taking place within a certain region, say, the inside of a gunpowder manufactory." Modern-day explosives manufacturers agree wholeheartedly. Brook, Moore, and Krider mention two American ordnance plants that blew up after lightning strikes in the 1970s.

The explosions took place under circumstances that kept the national news media away. When asked for more details, Moore, if he trusts his questioner, will give specific facts. Along with the facts he will give a brief lecture about earnest safety officers conscientiously applying what was then the state of the art in lightning protection.

The task facing these safety officers is formidable. Electric companies, for example, have to take measures to protect their equipment against lightning damage. But they use what is called "probabilistic protection." In other words, they take whatever steps they deem sufficient to keep most of the power on most of the time. In deciding, they balance the danger of consumer irritation against the high cost of lightning protection.

As Brook pointed out in the report, the foundation of probabilistic protection is indicated by the name: fuzzy estimates of the probability that lightning will strike the power grid in a certain region. When the stakes are fairly low— some equipment damage, a few individuals complaining because their freezers defrosted, and some time spent in turning the power back on—probabilistic protection is adequate. But in an ordnance plant, it is not good enough at all. If lightning gets into an explosives magazine, it is small comfort indeed to know that the odds were a million to one against it.

When the study began, the three researchers already knew one way to protect explosives plants from lightning. It is simple in theory: put the plant inside a Faraday cage, a completely enclosed and well-grounded metal structure. Maxwell recommended that approach in 1876. But putting a modern factory inside a perfect Faraday cage would be impossible; the electromagnetic pulse of a lightning stroke, for example, can induce a surge on power lines, giving lightning a path into an otherwise well-protected facility. It does not even take a direct hit.

Protection against such damage is the factory's responsibility. The equipment installed by the power company is there to protect the power company; it does very little for the consumer. Designers of lightning protection systems have to take a thousand such things into consideration.

In searching for ways to protect a factory for lightning, they start at the top. Lightning rods, invented by Benjamim Franklin, remain the basic instrument of protection from lightning damage. Franklin himself came up with two very different theories of how the rods might work.

Franklin invented the lightning rod in 1750 after investigating the point-discharge phenomenon, which enabled him to bleed away static charges with a sharp needle. Reasoning that the same principle should apply to charged thunderclouds, he put up sharp lightning rods to discharge the clouds slowly and safely. The idea is still current; a commercial lightning eliminator resembling a knight's mace was tested at Langmuir in 1981. The device was ruined by a direct hit, discharging jokes about the lightning eliminator that was eliminated by lightning.

In light of what is now known about lightning, such devices seem ridiculous. Lightning is a large-scale phenomenon that originates many kilometers above the ground, whereas point discharge is strictly a small-scale effect. But this was unknown in 1750; Franklin can hardly be faulted for his erroneous explanation of a valuable invention. He made observations and found out that trying to discharge the clouds would not work. Then he proposed, correctly, that the rods were useful because they intercepted lightning strokes and carried them harmlessly to ground.

But what is the optimum design for a lightning rod? Sharp

ones—the kind advocated by Franklin—and blunt ones are available. Both are supposed to intercept lightning, but evidence suggests that sharp rods often fail to do so. Sharp rods on the Apollo 12 gantry, on the Langmuir radar tower, and on one ill-fated explosives factory were bypassed by lightning in favor of lower objects supposedly within the rods' "cone of protection."

It seems that the sharp rods give off such a high point-discharge current that the electric field immediately above the tip is weakened, whereupon the lightning goes to something below that has a stronger electric field around it. The cone of protection supposedly extends downward and outward at a forty-five-degree angle from the tip of a lightning rod. Diagrams have been published showing the cone of protection below such things as radio antennas on boats. But, as Moore pointed out in the January 1983 issue of the *Journal of the Franklin Institute,* the cone of protection is not something to trust one's life to.

Besides the protection of buildings and aircraft, other, more esoteric problems appear in unexpected areas. For example, the spinning rotors of a helicopter can, under certain circumstances, build up enough electric charge to spell trouble for anyone who grabs the cargo hook. This has nothing to do with lightning, but it is an electrical discharge phenomenon, so Brook is currently investigating it.

Also of concern to the military is the chance that a ground-based particle-beam weapon, something that might be used if the Strategic Defense Initiative or "Star Wars" plan ever comes to fruition, could trigger lightning onto itself by creating an ionized channel to a cloudbase. High-voltage static electricity in all its forms is quirky enough—and dangerous enough—to give plenty of job security to those who would design ways of protecting against it.

Where Do They Go From Here?

Now that Langmuir Laboratory has been around for twenty-five years, it is natural to ask what the future holds in store. Of course, it's not possible to accurately predict the future of a course of investigation. But although the Langmuir researchers may not know exactly where they will end up, they do have a good idea of the direction they are going in.

First, they are going to keep looking for money. In 1935, E. J. Workman began his research efforts on a budget of about $350. Today, a single research project might cost in the neighborhood of $100,000 per year. A senior research physicist at a university laboratory draws an annual salary in the range of $30,000 to $40,000. The equipment in the scientist's lab is far more sophisticated than that of yesteryear—and far more expensive. The administrative overhead that Workman refused to pay to UNM is now accepted as a fact of university life. Even the physical necessities of running a remote laboratory—electricity, gasoline, food—cost much more than they did fifty or even twenty-five years ago. Today, more than ever before, scientific research is big business.

The primary support for Langmuir Laboratory has always come from the Office of Naval Research and the National Science Foundation in the form of grants both to the lab and to individual research projects. But the ONR, which is a branch of the Navy, and the NSF, which is the Federal government's principal scientific funding agency, are both subject to the budgetary whims of the president and the congress. The last few administrations have not been kind to basic research. Moore has his reservations about defense research, but both money and data are in store for the lab if weapons scientists come there. The NSF still provides a large part of the funding, but Langmuir scientists, like university researchers everywhere, are encouraged to find money wherever it is available.

But this need to search for money is taken for granted in the research world. When the scientists at Langmuir make plans, they think more in terms of unsolved problems in the field and the possible ways of investigating them.

In 1961, Marx Brook pointed out the usefulness of visual observation in an age of radars, remote sensors, and digital data-recording devices. His comments are as valid today as they were a quarter of a century ago. Fittingly, Brook is one of the leaders in using a new vantage point for observing lightning: the space shuttle.

Astronauts have always been interested in lightning, especially during countdown and launch, when it presents a hazard to their spacecraft and themselves. In orbit, far above the earth's atmosphere, they can take a more objective view. John Glenn, orbiting in his Mercury capsule, noticed lightning but did not pay much attention to it. Pete Conrad, returning from the moon aboard Apollo 12, reported seeing "lightning flashes ripping up the entire nighttime side of Earth." Edward Gibson, observing from Skylab, added a new twist. He reported that one bolt would seem to be followed by a great many more, almost as though there were some kind of collective organization to lightning strokes throughout a storm system.

For years, these remained isolated observations by nonspecialists, the kind of anecdotal data that does not carry much weight. For all their lack of scientific rigor, though, the astronauts' reports are enormously exciting to atmospheric physicists. Before the shuttle, it was impossible to study the subject further, because the astronauts had many other duties. But space flight is now somewhat more routine, so scientists can follow this line of inquiry.

That came to pass in 1981 with the introduction of the space shuttle. Brook, Vonnegut, and NASA's Otha H. Vaughan, Jr., developed lightning-observation equipment and personally trained several shuttle crews in lightning observation techniques: where to look, what they could expect to see, and what information was needed to make their observations scientifically useful. The second, fourth, and sixth shuttle missions carried optical detectors to spot lightning and sixteen-millimeter cameras with which to film it. Even these early efforts provided useful data. For example, lightning strokes were seen propagating horizontally at up to 320,000 feet per second and extending as far as fifty miles.

Complicating the picture, shuttle astronaut R. H. Truly reported that two thunderstorms above the Amazon appeared to be "talking to one another" in terms of simultaneous lightning. The scientists continue to entreat NASA to put lightning observations on the agenda for shuttle flights and to incorporate a lightning sensor in a geosynchronous satellite sometime in the late eighties. In the meantime, they wonder what might cause lightning flashes to be collectively organized, if indeed they are, and what old hypotheses this might support or disprove.

Now and then, NASA has also provided the Langmuir researchers with another observation platform, the U-2 reconnaissance plane. The U-2, designed for spy missions, is essentially a jet-propelled sailplane. At its cruising altitude of 70,000 feet, it can look down on an entire thundercloud, giving scientists a perspective in between the global panorama from orbit and the close-up view available from ordinary aircraft.

The question of cloud electrification, one of the fields

Bernard Vonnegut. Photo courtesy of Charles Moore.

Workman and Holzer probed in the mid-1930s, has yet to be conclusively answered. Moore, Vonnegut, and a few other atmospheric physicists believe that convective transfer of space charge plays the major role. Most of their colleagues disagree, but that does not stop communication on the subject; if anything, it helps the investigation by encouraging debate. The feelings run strong but not hard.

Of course, these scientists do enjoy their arguments. Back in the mid-fifties, Workman, Brook, and Reynolds would often gather in a downstairs lab beneath Workman's apartment in the Research Building for an evening of shoptalk. When the night grew short and the decibel level grew high, Workman's wife would bang on the floor with her cane. "You boys stop fighting," she would say. "It's time to go to bed."

Those days are long gone, and so are most of the people. Workman died in 1981 at his retirement home in Santa Barbara, California, having founded not one but two atmospheric research laboratories in his eighty-two years. Reynolds left both New Mexico Tech and atmospheric physics in 1955 to become the New Mexico State Engineer. But Brook is still there; he and Moore carry on a tradition as old as science itself, the tradition of hollering at each other over the divergent conclusions they have drawn from data available to both.

As the director of the Research and Development Division, Brook is Moore's boss as far as research is concerned. In an academic research setting, the term "boss" carries few of its industrial overtones, but more than once Brook could have probably driven Moore away from the school. It certainly would not have been the first time a senior professor had done such a thing. But Brook has chosen not to do so. High priest and heretic, both full professors, have their offices almost within shouting distance of each other in

Workman Center, and their differing ideas coexist in the *Journal of Geophysical Research.*

Perhaps one hypothesis or the other will win out; perhaps one man or both will live to see that day. It is also possible that some new hypothesis will displace both. Only time and experimentation will tell. The way in which these scientists work is very human, affected by emotion, politics, finance, and a thousand other seemingly unscientific factors. But the work itself is physics, and the ultimate test of a hypothesis is how well it explains some aspect of the physical world.

Toward that end, the charge-carrying cable across Sawmill Canyon, blown down in an ice storm in 1981, was put back up in 1984. Attempts to invert the polarity of clouds by releasing negative charge into their bases were apparently successful in 1984 and 1985, providing highly suggestive, partial evidence. Moore and Vonnegut are doing further work with the cable, trying to obtain proof, or at least the closest thing to proof that one gets in atmospheric physics.

While the atmospheric electricians, as they call themselves, wrangle over lightning and charge separation, others at Langmuir are studying the physical dynamics of the thundercloud. The National Center for Atmospheric Research and the National Oceanic and Atmospheric Administration provided four sophisticated Doppler radars in 1984. One radar was hauled up to the mountaintop. The other three were placed in a rough triangle on the surrounding plains.

The Doppler radars make use of the Doppler effect: the fact that the length of any wave, electromagnetic or mechanical, becomes longer or shorter depending on whether the source is going away from or toward the observer. This principle explains the rise and fall of the whistle of a passing train. The Doppler effect is used in various forms by a great many people. Astronomers measure the Doppler shift, or "red shift," of light from stars and galaxies to determine the motion of those heavenly bodies relative to the earth. State troopers use Doppler radar to measure the red shifts of receding sports cars.

In Doppler radar, the reflected beam is longer in wavelength that the transmitted beam if the target is receding. Similarly, it is shorter in wavelength if the target is approaching. By measuring the amount of change, atmospheric scientists can tell approximately what the winds in the different parts of a cloud are like. Doppler radars for atmospheric physics, equipped with high-speed data-recording equipment, cost upwards of a million dollars apiece. Langmuir Lab gets them now and then by inviting their federally funded owners to spend the summer at the lab.

The flow of air outside the actual storm is also important in these studies, but radars cannot "see" clear air no matter how fast it is moving. Aircraft, often including a Sabreliner business jet from the National Center for Atmospheric Research, are flown around the storms to measure these winds directly. The Sabreliner has also been used to drop radar-reflecting chaff into the clear air around the storms.

While most of the studies at Langmuir and the related research at New Mexico Tech have been pure science, directed toward a better understanding of thunderstorms, the work has some practical applications. In the early eighties, Brook helped develop a new kind of electrostatic precipitator to clean the fly ash out of smokestack emissions. By using an electrified aerosol of water droplets—in other words, applied cloud physics—this precipitator can scrub out tiny particles that conventional precipitators miss.

Although the new smokestack scrubber is applied science and has little to do with improving our understanding of clouds, it nonetheless reflects the essence of the way these physicists work. Mathematical ability is necessary to their work, of course, and so is a good understanding of instru-

Bill Winn in balloon hangar, 1977. New Mexico Tech Information Services.

mentation. But what really distinguishes them is the ability to gather all kinds of knowledge and skills and bring it all to bear on one narrow, specialized problem. No law says that the narrow problem has to be in one's own subspecialty; in fact, over a thirty-year career, a physicist is quite likely to probe several areas. Brook's venture into pollution control is no more unusual than Moore's teaching of a machine-shop course or Holmes's early work in geophysics.

There are other practical applications, both obvious and hidden, of the work being done at Langmuir. Studies of lightning and its physical and electromagnetic effects will undoubtedly continue to be one of the lab's main concerns. With two kivas built and running, the ground under the chicken wire on top of Baldy is becoming crowded. But two kivas bring in data that amounts to an embarassment of riches; as with many aspects of modern science, the problem is not to get the raw facts, but to get the time and personnel to put it all together and make sense out of it.

As the Irving Langmuir Laboratory for Atmospheric Research begins its second quarter-century, the various fields of inquiry related to thunderstorms will receive more and more attention. Many things have changed over the years. Old researchers have retired and new ones have begun working their way up. The facilities have been greatly expanded. Tube-type equipment that belongs in a museum is being replaced with ultramodern, solid-state, computerized apparatus that often does a better job. The scientists have made some important discoveries and have also added to the great ash heaps of negative data that result from most scientific research. But for all the changes, the primary goal remains the same: to apply the capabilities of modern instrumentation and the rigor of mathematical analysis to the study of one of the oldest enigmas on the planet, the thunderstorm.

Winter on South Baldy. Marvin Wilkening photographs.

APPENDIX I

The Principal Researchers Today

Marx Brook retired in early 1986 from the directorship of the Research and Development Division and from his professorship in physics at New Mexico Tech.

Jack Cobb retired as caretaker at Langmuir Laboratory in 1984, but still feeds the squirrels there and teaches physics majors how to check rain gauges and drive bulldozers.

Stirling Colgate is still in New Mexico as a Senior Fellow in the Theoretical Division at Los Alamos National Laboratory. He remains involved with research at Langmuir through an adjunct professorship and through running the Digitized Astronomy Telescope on South Baldy.

Charles Holmes continues as a professor of physics at New Mexico Tech and as a researcher at Langmuir Lab. He says he is "taking it one year at a time."

Robert Holzer is a professor emeritus at the Institute of Geophysics and Planetary Physics of the University of California at Los Angeles.

Lamar Kempton directs the Terminal Effects Research and Analysis (TERA) project, an explosives and ordnance research organization, at New Mexico Tech.

Paul Krehbiel is a senior engineer and research associate in New Mexico Tech's Research and Development Division.

Irving Langmuir worked as a research scientist and as a consultant at General Electric from 1909 until his death in 1957. The winner of the 1932 Nobel Prize in chemistry recently had a scientific journal, Langmuir, named in his honor.

Charles Moore retired from his physics professorship at New Mexico Tech and gave up the chairmanship of Langmuir Lab in late 1985. He continues as a principal investigator of atmospheric physics at the lab.

Steve Reynolds serves as the New Mexico State Engineer, a position he has held for more than thirty years.

Vincent Schaefer is director emeritus of the Atmospheric Sciences Research Center at the State University of New York in Albany. He recently co-authored the popular book, A Field Guide to the Atmosphere.

Kenny Sorensen retired from TERA in 1983 and lives in Socorro.

Bernard Vonnegut, although officially retired, still conducts research in cloud physics and thunderstorm electricity at the State University of New York in Albany. He expects to continue working at Langmuir Laboratory during the summers.

Marvin Wilkening retired as a professor of physics and dean of graduate studies at New Mexico Tech in 1984. He continues pursuing research in environmental radiation.

William Winn is the newly appointed chairman of Langmuir Laboratory as well as a professor of physics at New Mexico Tech.

E. J. Workman, after retiring from New Mexico Tech in 1964, went to Hawaii to help establish the Cloud Physics Observatory at the University of Hawaii-Hilo. He retired again in 1970 to Santa Barbara, California, where he died an octogenarian in 1982.

APPENDIX II

The Students Of TC 301

Jill Bartel is an associate writer/editor with Martin-Marietta Aerospace in Orlando, Florida.

Chris Benedict is a science fiction writer living in Las Vegas, Nevada, with her husband.

Joe Chew, a technical writer for a San Francisco computer firm, is collaborating with Dr. Jimmie Killingsworth of Texas Tech and Dr. Michael Gilbertson of New Mexico Tech on a technical-writing textbook.

Rick Clyne completed an internship with ANORAD Company in Hauppauge, New York, in the summer of 1985, and continues to work on his technical communication degree at New Mexico Tech.

Kim Eiland, a writer/editor in the Information Services Office at New Mexico Tech, received the bachelor of science in technical communication with highest honors from Tech in May 1986.

Diane Hattler is a technical editor for the U.S. Geological Survey in Fairfax, Virginia.

Terry Jackson, a technical writer for Science Applications International Corporation, recently organized the Southern Nevada Chapter of the Society for Technical Communication and served as the chapter's first president.

Toni Ball Johnson, winner of the 1985 Sylvester Award as New Mexico Tech's top humanities graduate, works at the school's Technological Innovations Center and does free-lance writing and editing.

Mary McClure holds a writing position with Burroughs in Orange County, California.

Heidi Miller lives in Los Lunas, New Mexico, and works in Albuquerque.

Dave Pellatz, the only one in the group who did not major in technical communication, works as a petroleum engineer for Conoco in Casper, Wyoming.

Kathy Smith, after completing a six-month cooperative internship at Los Alamos National Laboratory, received the bachelor of science in technical communication with high honors from New Mexico Tech in May 1986. She now lives in Idaho.

Bibliography

In addition to the references cited below, other sources for this study were interviews, newspaper articles, and unpublished archival material. The interviews involved many people who worked at Langmuir; cassette tapes of interviews with Marx Brook, Charles Holmes, Lamar Kempton, Charles Moore, Steve Reynolds, Kenny Sorensen, and Bernard Vonnegut. In an effort to cross-check the accuracy of this oral history, the manuscript was read completely or in part at various stages by Marx Brook, Charles Moore, Bernard Vonnegut, and Bill Winn.

Hundreds of newspaper articles were read in the preparation of this book. Among the newspapers used were the following: Albuquerque Journal, Albuquerque News, Albuquerque Tribune, Alumni Bulletin (New Mexico Tech), Asian Student, Carlsbad Current-Argus, Clayton News, Datil Roundup, Denver Post, El Arrastre (former New Mexico Tech student newspaper), El Paso Herald-Post, El Paso Times, Gold Pan (New Mexico Tech alumni newspaper), Hobbs News-Sun, Las Cruces Sun-News, Lincoln County News, Los Alamos Monitor, Lovington Daily Leader, Magdalena Roundup, New York Herald Tribune, New York Times, Roswell Daily Record, Roy Record, Santa Fe New Mexican, Silver City Daily, Silver City Enterprise, Socorro Defensor Chieftain, Taos El Crepusculo, Tucumcari American, and USA Today.

Archival materials included New Mexico Tech's administrative files for the years 1935–1985, a large body of letters from the E. J. Workman administration, a collection of periodic research reports to the U.S. Army Signal Corps, an informative letter from Robert Holzer, and the transcript of a talk in which E. J. Workman recalled the early years of his career.

Those who want access to these materials should contact the archivist at the Martin Speare Memorial Library of New Mexico Tech.

Marx Brook's archive file, kept by the Research and Development Division Photo Lab, was the principal source of historical photographs. The file contains four-by-five-inch, two-and-a-quarter-inch, and thirty-five-millimeter negatives dating back to 1946. Color negatives and slides and black-and-white prints are also stored in that file. Charles Moore, Marvin Wilkening, and Bill Winn also maintain large individual collections of slides and photographs pertaining to atmospheric research and Langmuir Laboratory.

Books

Frazier, Kendrick. The Violent Face of Nature. New York: William Morrow and Company, Inc., 1979.

Moore, C. B., and B. Vonnegut. "The Thundercloud," in Lightning, edited by R. H. Golde. New York: Academic, 1977.

Moore, C. B., B. Vonnegut, and A. T. Botka. "Results of an Experiment to Determine Initial Precedence of Organized Electrification and Precipitation in Thunderstorms," in Recent Advances in Atmospheric Electricity. London: Pergamon Press, 1958–59.

Schaefer, V. J., and J. A. Day. A Field Guide to the Atmosphere. Boston: Houghton Mifflin Company, 1981.

Uman, Martin A. Understanding Lightning. Pennsylvania: Bek Technical Publications, Inc., 1971.

Wallis, W. A., and H. V. Roberts. Statistics: A New Approach. Glencoe, Illinois: The Free Press, 1956.